Hydrogen in an International Context: Vulnerabilities of Hydrogen Energy in Emerging Markets

RIVER PUBLISHERS SERIES IN RENEWABLE ENERGY

Series Editor

ERIC JOHNSON
Atlantic Consulting
Switzerland

The "River Publishers Series in Renewable Energy" is a series of comprehensive academic and professional books which focus on theory and applications in renewable energy and sustainable energy solutions. The series will serve as a multi-disciplinary resource linking renewable energy with society. The book series fulfils the rapidly growing worldwide interest in energy solutions. It covers all fields of renewable energy and their possible applications will be addressed not only from a technical point of view, but also from economic, social, political, and financial aspect.

Books published in the series include research monographs, edited volumes, handbooks and textbooks. The books provide professionals, researchers, educators, and advanced students in the field with an invaluable insight into the latest research and developments.

Topics covered in the series include, but are by no means restricted to the following:

- Renewable energy
- Energy Solutions
- Energy storage
- Sustainability
- Green technology

For a list of other books in this series, visit www.riverpublishers.com

Hydrogen in an International Context: Vulnerabilities of Hydrogen Energy in Emerging Markets

Editor

Ioan Iordache

National Research and Development Institute for Cryogenics
and Isotopic Technologies ICSI,
Rm. Valcea, Romania
and
Romanian Association for Hydrogen Energy,
Rm. Valcea, Romania

Published, sold and distributed by:
River Publishers
Alsbjergvej 10
9260 Gistrup
Denmark

River Publishers
Lange Geer 44
2611 PW Delft
The Netherlands

Tel.: +45369953197
www.riverpublishers.com

ISBN: 978-87-93379-98-5 (Hardback)
978-87-93379-99-2 (Ebook)

Contents

PART II: Hydrogen Energy in Emerging Markets

Foreword

As I write, the referendum in Great Britain took place and the result took many people by surprise. The result is a divided country, geographically and demographically, where Scotland and Northern Ireland wanted to remain in Europe, Wales, and England voted to leave the Union. The majority of young people voted to remain in Europe as they believed that this was the best for their future. This referendum painfully showed us that the European Union must look for a new project; a project where each European feels involved and is worried about. Climate change could be such a topic. Many people are concerned about it no matter what their religion, age, social status, or political conviction is. Climate change is an item that does not stop at country-borders and forces countries to work together. To decarbonize the society, an energy transition from fossil fuels to renewables is inevitable. Sun, wind, and hydro will all play a role in this required transition. The need to transport the energy to the users and to store it in case of excess energy will become imminent in the future.

Against this background, hydrogen is one of the most abundant elements on earth and is ideal as an energy carrier. The technologies related to hydrogen such as fuel cells and electrolyzers are advancing everywhere in the world. Today, we can see the first products like cars, buses, and CHP systems entering the European market. It must be noted that many of these products today are still expensive comparing to the regular, fossil fuel-based technologies, but it is clear that the gap is slowly closing. The advancement of the sector related to hydrogen technologies is partly due to the Fuel Cells and Hydrogen Joint Undertaking (FCH JU), a public private partnership that was set up between the European Commission, the industry, and the research institutes. This unique partnership is speeding up the development of the different technologies and bringing them faster to the market. Looking into participation of beneficiaries from the new Eastern European Member States, we see that it has been steadily increasing over the past years. The main interest of the Eastern European participants lies in the field of stationary power production and combined heat and power production and is mainly concentrated in universities and research centers with a limited participation of industries or small and medium enterprises. On one hand, this clearly highlights the high level of the universities and research institutes in Eastern Europe, bringing a clear added value to the research performed in the rest of Europe. On the other hand, it also indicates that the industrial network in these countries is not yet ready to participate in a transition toward new clean energy and emissions-free transport processes. Nevertheless in these countries, it is interesting to note that the actors active in the deployment of fuel cell and hydrogen technologies start to build national coalitions or societies and that in some niche markets the related industries start to be at the lead in Europe, such as manufacturers of hydrogen buses or fuel cell-powered electric small airplanes. The book will demonstrate the status of the technology and the opportunities in vulnerable energy markets and emerging countries.

Bart Biebuyck
Executive Director FCH2 JU

Executive Summary

The book contains plenty of examples about hydrogen research and development progress in different countries. Each chapter delineates a specific situation for a country and one of them describes some aspects at EU level. The examples described into the book are not aleatory. There are examples for countries that are leaders at European or international level, for example, Germany, South Korea. For these chapters, the intention of authors, editors, and publishers are to give the reader possibility to understand the concrete examples about what means the "success story" of the hydrogen. The chapter dedicated to South Africa offers other specific lessons, the country has a huge potential on a specific market, platinum and palladium utilization for hydrogen and fuel cell technology but, in the same time, there is a significant demand for energy. The chapters dedicated to The Czech Republic, Poland, and Romania reveal the commitment of Eastern European countries in this adventure often viewed today as a subject of very advanced countries. The specific situation in Russia Federation describes a strong background, an uncertain present, and a questionable future for the hydrogen and fuel cell technology. Finally, the EU, by voice of the main stakeholders, considered the hydrogen and fuel cell a decisive issue, with economic and societal ramifications.

The start-up phase of the basic research began in **Germany** in the 70's. The phenomenon was set in motion by the energy crisis. The main effort to overcome that crisis was in research for alternative, sustainable fuels for transportation, especially. The research program, funded by the Federal Ministry for Research and Technology, approached the general public with the publication "Neuen Kraftstoffen auf der Spur" (The way to new fuels) in 1974. The possibility to produce hydrogen via electrolysis of water using renewable energies made this energy carrier attractive. In the case of hydrogen, a lot of basic research had to be done concerning storage, safety, mixture formation combustion. It was carried out in the priority project "Alternative Energy Sources for Road Transport", which was initiated and sponsored by the Federal Ministry for Research and Technology in 1979. Five vehicles using mixed hydrogen/gasoline and five powered by hydrogen only were planned to be tested in Berlin city traffic by selected operators from October 1984 to March 1988. The hydrogen fleet operation tests required a lot of the basic research and development. In addition to classical compressed or liquid hydrogen storage, at the beginning of the 70's, hydrogen storage by special metal hydrides was discovered, and the development of the metal hydride containers for automotive, stationary, and submarine applications was initiated. The interest of the German industry in renewable and alternative energy systems decreased because of the ample supply of crude oil, and due to the fact that exhaust gas emission problems were intermediately solved by catalysts at the end of 80's.

Renewed concerns about the future energy supply with respect to the limited fossil energy resources and CO_2 emissions gave a new momentum to the hydrogen energy research and development. The cooperation between Germany and Canada under the EU umbrella led to "Euro-Québec-Hydro-Hydrogen Pilot-Project (EQHHPP)". The project investigated how and under which conditions renewable primary energy in the form of hydroelectricity can be converted via electrolysis into hydrogen. Theoretically, they investigated a hydrogen transport corridor between Québec and Europe. The hydrogen would be shipped, stored, and used in different ways: electricity/heat co-generation, vehicle, aviation, ship propulsion, and steel fabrication. The original plans intended to include for the project are the following: hydrogen production and storage, transoceanic hydrogen transportation, use of hydrogen in a bus fleet, cogeneration plants, and an Airbus turbine. The development of hydrogen technology with PEM fuel cells was mainly promoted by the automotive industry. The demonstrative projects for passenger cars, buses, and special mobile applications were implemented by the National Organization for Hydrogen and Fuel Cell Technology (NOW) under the National Hydrogen and Fuel Cell Technology Innovation Programme (NIP). In Germany not only transportation needs are to be partially covered by hydrogen and fuel cell, nearly 30% of primary energy consumption in Germany are used in households. This energy can be partially based on hydrogen and fuel cell as well. Here, the German industry cooperates at the national, European, and international level to improve the efficiency of energy utilization, in the transition phase even with natural gas utilization in SOFC fuel cell and PEM fuel cell via reforming.

The volatility of primary energy sources, especially renewables, is a major challenge for the future sustainable energy systems. Energy storage will be an indispensable factor for the basic energy supply system. With respect to energy density, the storage of chemical energy (enthalpy of chemical reaction) has much higher values than the storage of potential energy. Hydrogen underground storage in salt caverns enables not only the necessary energy storage for stabilization of the electric grid but also the supply of hydrogen for other applications.

The joint initiative of the government and industry, along with cooperation among research institutions, industrial companies, and public institutions, is the key solution for expansion of the hydrogen infrastructure in both transportation and stationary applications in Germany.

Because of its troubled history, **Republic of Korea** (South Korea) was forced to experiment all fuels available, wood, coal, oil, and natural gas, in the second part of 20th century. But for the 21st century, this country want to turn for ever at the final chemical fuel, the hydrogen and its most efficient engine, the fuel cell. Starting with 80's, the researchers and scientists from South Korea have done first pioneer studies and actions regarding utilization of hydrogen for energy purposes, Hydrogen and Fuel Cell Laboratory in Chonnam National University is an example of growing interests for hydrogen energy and fuel cell vehicles. The scientifically community keep in touch and closed cooperation with international community of hydrogen, and them constantly presence in the actions under auspices of International Association for Hydrogen Energy is a proof of gratitude for the effort made Korean hydrogen community. The colleagues from Chonnam National University in cooperation with other hydrogen stakeholders were able to organize the 20th World Hydrogen Energy Conference in Gwangju city and the selected theme of it was "Creating a healthy and peaceful hydrogen energy world".

The public support of the Korean government has started since 1988 by both Ministry of Commerce, Industry and Energy (MOCIE) and Ministry of Science and Technology (MOST). Thereafter, the finances have become specialized on specific organizations and programs, National RD&D Organization for Hydrogen and Fuel Cell, and 21st Frontier Hydrogen Energy R&D Program.

The hydrogen utilization for mobility had the same evolution as worldwide, initially it was used as fuel in internal combustion engine and gas turbine and finished by being used in fuel cells. The Korean automotive industry had begun the testing of hydrogen end fuel cells vehicle relatively late, in 2000's. But with much toil and effort, they were able to have the world's first mass production of ix35 (Tucson) fuel cell electric vehicles in 2013. Most of the 17 hydrogen stations were built by support of the government from 2001 to 2014.

The residential utilization of hydrogen energy is other sectors where the Koreans make efforts. Ministry of Trade, Industry and Energy (MOTIE) and Ulsan City Government have been supporting Ulsan Hydrogen Town from 2012 to 2018. A demonstrative number of apartments, houses, or offices use hydrogen and fuel cells. The MOST has been supporting by a large amount budget to Korea Atomic Energy Research Institute (KAERI) to produce hydrogen by nuclear energy with Very High Temperature gas-cooled Reactor (VHTR) for 16 years, from 2004 to 2020.

Finally, but not at the least, Human resource in Korea is very much important and many professors, students, researchers, and officials are working for the hydrogen and fuel cells in various universities, companies, many local governments as well as central government and institutes.

The evolution of hydrogen and fuel cell research in The **Czech Republic** follows a similar line of all former Soviet bloc countries. First systematic research was related to the nickel and KOH-based systems in seventies of 20th century. This start was abandoned relatively soon because of limited resources available especially for the fuel cells components to be imported from outside Soviet bloc and because under communist ideology the technological development was often limited to the optimization of the existing approaches over the introducing innovative, breakthrough technologies. Restructuring of the industry and economy after the collapse of the Soviet bloc pushed somewhat the research and development stakeholders from The Czech Republic to search for new domains of research allowing building up on the existing competences. A number of subjects have become interested in hydrogen and fuel cell: University of Chemistry and Technology Prague (UCTP), Institute of Macromolecular Chemistry of the Czech Academy of Sciences of The Czech Republic (IMC CAS), J. Heyrovský Institute of Physical Chemistry of the Academy of Sciences, Czech Technical University in Prague, Technical University of Liberec, Technical University of Ostrava, ÚJV Řež, Faculty of mathematics and physics, Charles University in Prague (FMF CU) and Astris. Czech Hydrogen Technology Platform (HYTEP) opened new period of hydrogen technologies in The Czech Republic and was established under auspice of the Ministry of Industry and Trade of The Czech Republic. HYTEP has chosen to organize an international conference named Hydrogen Days, it is becoming more famous from one edition to another.

The Czech Republic has successfully finalized a series of projects: TriHyBus, Hydrogen filling station, Neratovice, Solid oxide steam electrolyzer

(SOSE), Autarkic system, Platinum free novel electrocatalyst, etc., and participated successfully in several European projects.

TriHyBus is an EU demonstration project led by ÚJV Řež. ThiHyBus is a fuel cell bus prototype based on 12m Irisbus Citelis design with the overall capacity 96 passengers, of which 26 only seated. The project story shows a simple lesson: even an ingenious technological idea brought into the demonstration project needs a powerful management with a strategic vision and a strong commitment, to maximize the project benefits and to ensure the project continuation.

In **Poland,** the diversification of energy sources and rational energy policy and security are a very significant issue. The current state of the Polish energy industry is strong reliance on bituminous and brown coal. There are relatively small deposits of natural gas and petroleum, which are mostly imported. The high efficiency of fuel cell technology would allow the existing deposits to be utilized to a much larger extent by and to obtain more energy from the same amount of resources. The relatively uniform distribution of resources across the country, which would allow the production of fuel for the fuel cells on the site, eliminating the need to develop new power transmission infrastructure. The gas and electrical energy transmission infrastructures are well-developed and there are numerous refineries, liquefied natural gas import terminals, underground gas storage facilities, and structures that allow the inexpensive production of hydrogen. However, the utilization of hydrogen as an energy carrier remains marginalized in the context of the future energy industry. The official documents completely omit the question of gradual transition from conventional power generation based on fossil fuels to a low-emission energy production using devices as fuel cells. Hydrogen-based energy generation can be subsumed under the term "technological diversification" included in the official document entitled "Energy Policy of Poland until 2030". The document emphasizes the need to implement modern technologies in energy sector and formulates the requirement to support the research and development of new technologies and solutions. There are certain common points between the Polish research and development policy and general energy policy.

There are several clusters of companies, research organizations, business-related organizations, and public entities that declare in their statutes and corporate charters that will cooperate in the field of renewable energy and hydrogen-based energy. Here are mentioned, among others, Polish Hydrogen and Fuel Cell Association and Polish Hydrogen and Fuel Cell Technological Platform. At the national level, there are 22 entities, universities, and institutes, where research teams are active in the field of hydrogen and fuel cell. Although there are a number of events and national institutions capable to promote and implement hydrogen economy, the lack of a consistent policy makes it impossible to do.

Romania has good opportunities to make the transition from the dependence on fossil fuels to a power industry based on diverse energy carriers (such as hydrogen, among others), and to a power sector utilizing hydrogen and renewable energy sources. The replacement of actual systems would offer the opportunity to implement fuel cell stacks and to use hybrid systems that combine conventional and modern methods of electricity generation. By adopting hydrogen-based technologies and thus reducing greenhouse gas emissions, the Romanian stakeholders would be able to make profits or to recover their investments via the so-called emissions trading.

In Romania, there are little more than 100 research institutes and universities with research laboratories and a part of them are involved in the field of hydrogen and fuel cell research, development, and innovation. The scientific activity covers diverse fields: the generation and storage of hydrogen, fuel cell construction and electric properties, mobility, etc. The projects conducted in Romanian are targeted more toward research and less on development or demonstration. The researchers are well inspired and focus on maximum use of the existing potential. The flagship example is participation in *HyUnder* project, where, in cooperation with others partners from Europe, was studied and demonstrated the potential of hydrogen underground storage into salt caverns. Another example is utilization of H_2S from the Black Sea area, it must be mentioned that the subject is actively addressed in several costal countries. Other researchers from Romania are turning on more defined themes such as membranes of palladium/ceramic, others are baked in multidisciplinary projects that relate to the production of hydrogen through the thermochemical Cu-Cl cycle, taking into account Romania's nuclear energy infrastructure. At the national level, the research, development, and innovation authorities have financed about 80 scientific small and medium projects starting with 2000.

The National Center for Hydrogen and Fuel Cell function has been working at full capacity and extending constantly its facilities since 2010. It is a good example regarding the recognition of the competence and capacities is it affiliation as full member in N.ERGHY group. Another entity that advocates for promotion of hydrogen and fuel cell technology is Romanian Association for Hydrogen Energy, organization active also at European and international level.

The positive previous enumeration is counterbalanced by the absence of a national program for hydrogen and fuel cell research, development, innovation, and demonstration. The hydrogen and fuel cells topic can be found under other research directions like energy, materials, or environmental protection. Although Romania is a country that traditionally produce hydrogen, there was identified 13 big industrial producers, it is used entirely as a raw material in the petrochemical industry, agrochemical, metallurgy, glass, food, etc.

Aforementioned indices that mainly interest for hydrogen and fuel cell technology are in universities and research centers with a limited participation of small, medium, or industries enterprises. This highlights the high level of the universities and research institutes and, on the other hand, it also indicates that the industry is not yet ready to participate in a transition toward new clean energy and emissions free transport processes.

Russia Federation, like many other countries, has a "history" of hydrogen. First, utilization of hydrogen as an alternative fuel took place during World War II but the first scientific approach about utilization of hydrogen as an alternative fuel was realized in 70's by several research institutes and universities. The main goal of hydrogen utilization, in different mixtures, was to decrease the toxic wastes, but the progress in the exhaust gas emission catalytic systems and the absence of refueling infrastructure have not attracted enough industry attention. Another landmark achievement is to develop of TU-155 aircraft with hydrogen engine. The aircraft used liquid hydrogen as fuel and duration of flight could reach two hours. The USSR space project has led to the development of alkaline fuel cells in 60's. The researches and development was concretized with 10 kW alkaline fuel cell named "Photon" for space shuttle "Energy-Buran" in 1988. In parallel have been developed alkaline fuel cells for buses and for submarines. Other types of fuel cell developed in the same period include PEM and SOFC. Before 2003, several experimental vehicle with alkaline fuel cells were built by the largest Russian car company.

The alkaline electrolyzers are produced in Russia starting with 1949. PEM experimental electrolyzers and high-temperature solid oxide electrolyzers were developed and successfully tested starting with the end of 70's. The catalytic and plasma-chemical production of hydrogen from natural gas and other organic fuels were developed by research groups in Russia also. In 1996, Russian–American consortium on solid oxide fuel cell was organized. The main activity was concentered on the development of SOFC 1–5 kW power plants. The consortium was operating about 8 years. Also the concept of hydrogen production through the high temperature nuclear reactors was developed by a series Russian Academicians starting with 70's.

In our days, the support for hydrogen energy research and development is reduced and more focused on the fundamental research. The main reason is the unfavorable financial situation and low oil price. Despite these constraints, some researches are still visible, for example, an experimental plant (100 kW) for hydrogen production from Al or a SOFC 5 kW power plant with tubular cells modules. At the end of this paragraph, it is worth to be mentioned that regular conferences occurring and scientific journals publish articles on the topic of hydrogen and fuel cell in Russia Federation, indeed mainly in Russian with brief English annotations.

The quantities of hydrogen produced for different industrial purposes is huge, 4.5 million tons, about 8% of the world production. The main consumer of hydrogen is chemical industry, especially ammonia and methanol, and petrochemical refining followed by metallurgy, glass, food, and electronic industries.

Earlier-mentioned phrases indices that hydrogen is used exclusive as raw material in different industries, but the big step from chemistry to energy, where hydrogen to be used as sustainable energy vector, will not be made so soon.

South Africa, through HySA (Hydrogen South Africa) programs, as well as through the business-driven approach of PGM (Platinum Group Metal) mining companies, are aggressively pursuing research, development, and market creation for hydrogen, fuel cell, and water electrolysis technologies. These activities will result in beneficiation of mineral resource, in turns increasing employment for high-quality jobs, development of human capital, and improved quality of life. In South Africa, there is a significant demand for energy in rural areas. This includes critical infrastructure sites such as schools and hospitals, as well as households. Extension of the electrical grid to most of these areas is currently not economically viable. This represents a great opportunity to fulfil the unmet needs for on-site power generation using clean energy technology approaches. Other market opportunities for hydrogen and fuel cell technology include the mining and the telecom sectors (especially in the whole of the sub-Saharan area).

In May 2007, Hydrogen South Africa (HySA) was initiated by the Department of Science and Technology (DST) of Republic of South Africa and approved by the Cabinet. HySA is a long-term (15-year) programme within their Research, Development, and Innovation (RDI) strategy, officially launched in September 2008. This National Flagship Programme is aimed at developing South African intellectual property, knowledge, human resources, products, components, and processes to support the South African participation in the nascent, but rapidly developing international platforms in Hydrogen and Fuel Cell Technologies. HySA comprises of three R&D Centers of Competence: HySA Catalysis, HySA Systems, and HySA Infrastructure.

Benefits of developing hydrogen infrastructure and fuel cell market are: means of meeting increasing demand for energy, reduction of carbon footprint, platform for mineral beneficiation, opportunity for job creation, wealth creation, export opportunities, and increase demand for Platinum Group Metals.

The **Europe Union** SET-Plan (Strategic Energy Technology Plan) establish seven technology focus areas: wind energy, solar energy, carbon capture and storage, nuclear energy, bio-energy, electricity grids, and also fuel cell and hydrogen. At the EU level, the support of hydrogen and fuel cell is very vocational from the political point of view and strong in financial terms, approx. 1,335 M Euro until in 2020. The main vehicle for promoting this issue is Fuel Cell and Hydrogen Joint Undertaking (FCH JU). It is a legal entity entrusted with the coordinated use and efficient management of the EU funds. FCH JU (FCH2 JU from 2014, under Horizon 2020) is a public-private partnership with three members: the European Union, represented by the European Commission; the industry grouping named Hydrogen Europe; and the N.ERGHY Research Grouping.

The FCH JU programme jointly contributed 50/50 by public and private partners, has served as a key growth catalyst for hydrogen and fuel cell in Europe. The programme united the various stakeholders in the sector and provided predictability. This long-term commitment offers a stable framework for the research, development, and demonstration activities, which otherwise would have been impossible in difficult economic times. The FCH JU was and is in the situation to put the individual players together into a successful manner and links between national initiatives. Through its Annual and Multiannual Implementation Plans, FCH JU developed a joint coordinated strategy to increase effectiveness. The FCH JU funding mechanism has pooling resources to support nascent technologies beyond local or private possibilities.

As a result of FCH JU support, the hydrogen and fuel cell sector has grown substantially: 10% average increase in annual turnover, 500 M Euro in 2012, 8% average increase of research and development expenditures, 1,800 M Euro in 2012, 6% average increase of market deployment expenditures, 600 M Euro in 2012, 6% growth in jobs per year while the EU job market has contracted, and 16% annual increase in patents. The most recent programme, 2008–2014, has provided a strong and stable growth platform for Small and Medium Enterprises (SMEs), who have considered valued partners of hydrogen and fuel cell community. The SME participation rate in hydrogen and fuel cell-related projects was 25.6%, higher than that FP7 average rate of 18%. The focus of FCH JU projects, achieving substantial progress in both energy and transport, ranges from basic research to large-scale demonstration and pre-market studies. The next paragraphs contain general information in regard to references on thematic of projects according with the same bibliographic source.

At the same time, there is need to avoid geopolitical and societal fragmentation, there is need to develop a series of action for the concatenation of non-active players with the potential for hydrogen community. EU will solve this issue by approaching next tree directions: to build-up an energy landscape adequate with the requests of our century, to actively contribute to mature the hydrogen and fuel cell technology, especially competitiveness and price, and the public policy support, especially money investment.

List of Contributors

Bruno G. Pollet, *Power and Water (KP2M Ltd), Swansea, SA6 8QR, Wales, UK*

Byeong Soo Oh, *Chonnam National University, 77 Yongbong-ro, Buk-gu, Gwangju, 61186, South Korea*

Dmitri Bessarabov, *DST HySA Infrastructure Center (Hydrogen South Africa), North-West University, Faculty of Engineering, Private Bag X6001, Potchefstroom, 2520, South Africa*

Ioan Iordache, *1) National Research and Development Institute for Cryogenics and Isotopic Technologies ICSI, Rm. Valcea, Romania*
2) Romanian Association for Hydrogen Energy, Rm. Valcea, Romania

Ioan Ştefănescu, *1) National Research and Development Institute for Cryogenics and Isotopic Technologies ICSI, Rm. Valcea, Romania*
2) Romanian Association for Hydrogen Energy, Rm. Valcea, Romania

Jakub Slavík, *Independent business consultant specialized in transportation and smart city projects, The Czech Republic*

Johannes Töpler, *German Hydrogen and Fuel Cell Association (DWV), Berlin, Germany*

Karel Bouzek, *Department of Inorganic Technology, University of Chemistry and Technology Prague, Technická 5, 166 28 Prague 6, The Czech Republic*

Karin Stehlík, *1) Centrum výzkumu Řež, s.r.o., Hlavní 130, 250 68 Husinec-Řež, The Czech Republic*
2) Czech Hydrogen Technology Platform, Hlavní 130, 250 68 Husinec-Řež, The Czech Republic

M. Stygar, *Faculty of Materials Science and Ceramics, Department of Physical Chemistry and Modeling, AGH University of Science and Technology, Kraków, Poland*

Martin Paidar, *Department of Inorganic Technology, University of Chemistry and Technology Prague, Technická 5, 166 28 Prague 6, The Czech Republic*

Sergey Grigoriev, *National Research University "Moscow Power Engineering Institute", Krasnokazarmennaya str., 14, Moscow, 111250, Russia*

T. Brylewski, *Faculty of Materials Science and Ceramics, Department of Physical Chemistry and Modeling, AGH University of Science and Technology, Kraków, Poland*

Vladimir Fateev, *National Research Center "Kurchatov Institute", Kurchatov sq., 1, Moscow, 123182, Russia*

List of Figures

List of Tables

List of Abbreviations

AfDB	African Development Bank
AGH-UST	AGH University of Science and Technology (also: University of Science and Technology)
AIP	Air Independent Propulsion
BRICS	Brazil, Russia, China and South Africa
CC	Colocation Center
CCS	Carbon capture and storage
CENT	Center of New Technologies at the University of Warsaw
CoC	Center of Competence
CSP	Concentrating Solar Power
CTL	Coal to Liquids
DST	Department of Science and Technology
DTI	Department of Trade and Industry
EHA	European Hydrogen Association
EIT	European Institute of Innovation and Technology
ETV	Zakład Weryfikacji Technologii Środowiskowych (eng. Environmental Technology Verification)
EU ETS	EU Emissions Trading System
EU	European Union
GT	GigaTonne
GVP	General Verification Protocol
GW	GigaWatt
HCD	Human Capital Development
HEI	High Education Institution
HFCT	Hydrogen and Fuel Cells Technologies
HVAC	Heating, Ventilation, Air Conditioning
HySA	Hydrogen South Africa
IATI	Institute of Highway Technology and Innovation
ICE	Internal Combustion Engine
IDC	Industrial Development Corporation
IEA	International Energy Agency

IRENA	International Renewable Energy Agency
IRP	Integrated Resource Plan
ISI Master Journal List	Institute for Scientific Information Master Journal List
ITP	Instytut Technologiczno-Przyrodniczy (eng. The Institute of Technology and Life Sciences)
JSC	Joint-Stock Company
JWTS	Jednostka Weryfikująca Technologie Środowiskowe (eng. Polish Environmental Technology Verification Body)
KIT	Karlsruhe Institute of Technology
LFL	Lower Flammable Limit
MW	MegaWatt
NECSA	Nuclear Energy Corporation of South Africa
NEP	Nuclear Energy Policy
NFOŚiGW	Narodowy Fundusz Ochrony Środowiska i Gospodarki Wodnej (eng. The National Fund for Environmental Protection and Water Management)
NRC	National Research Center
NWU	North-West University
P2G	Power to Gas
PAFC	Phosphoric Acid Fuel Cell
PAS	Polish Academy of Sciences
PATH	Partnership for Advancing the Transition to Hydrogen
PEMFC	Proton Exchange Membrane Fuel Cell
PEMFC	Polymer-Membrane Fuel-Cell
PGM	Platinum Group Metals
PHAFCA/ PSWiOP	Polskie Stowarzyszenie Wodoru i Ogniw Paliwowych (eng. Polish Hydrogen and Fuel Cell Association)
PSA	Pressure-Swing Adsorption
PV	PhotoVoltaic
RAS	Russian Academy of Sciences
RSA	Republic of South Africa
RTD	Research and technology development
SEED	Sustainable Energy and Environment Development (name of the conference)
SEZ	Special Economic Zone
SOFC	Solid oxide fuel cell, also: IT-SOFC – Intermediate Solid oxide fuel cell

SSA	Sub-Saharan Africa
STN	Studenckie Towarzystwo Naukowe (eng. Students' Research Association)
UCT	University of Cape Town
URE	Urząd Regulacji Energetyki (eng. Energy Regulatory Office)
UWC	University of the Western Cape

PART I

International Success Stories for Hydrogen

1

Hydrogen Technology and Economy in Germany—History and the Present State

Johannes Töpler

German Hydrogen and Fuel Cell Association (DWV), Berlin, Germany

1.1 Introduction

The history of the development of hydrogen energy technology in the Federal Republic of Germany began in the mid-seventies of the past century. This process was mainly influenced by the energy crisis in this time, which resulted in Germany even in some "car-free Sundays" (Sunday with completely forbidden road traffic).

But the energy crisis was mainly a crisis of the availability of mineral oil so that the main focus of efforts to overcome this crisis was in research for alternative and mainly sustainable fuels, especially for automotive applications.

The first study how to reach this goal was "Neuen Kraftstoffen auf der Spur" (The way to new fuels) [1] published in 1974 within the framework of a research program sponsored by the Federal Ministry for Research and Technology. This study concentrated on methanol and hydrogen as alternative fuels and documented that hydrogen was on the verge of gaining new significance as a future alternative and sustainable fuel.

Simultaneous activities worldwide such as the foundation of the "International Association for Hydrogen Energy" (IAHE) confirmed this development.

The possibility to produce hydrogen via electrolysis from renewable energies and to use it without polluting the environment make this energy carrier so attractive.

In order to coordinate existing fundamental research activities in alternative fuels and to transfer it to practical testing programs, alcohol-, electric- and hydrogen-drives were investigated in the priority project "Alternative

Energy Sources for Road Transport" from 1979. This project was initiated and sponsored by the Federal Ministry for Research and Technology. In the special case of hydrogen, a lot of basic research work had to be done with respect to storage, safety, mixture formation, combustion, and so on.

Parallel to this development project, in 1978 the German government decided to support a vehicle-testing and demonstration project "Alternative Energy Sources for Road Transport" to be carried out in West Berlin. For this project, five vehicles for mixed hydrogen/gasoline operation and five for hydrogen operation only were planned to be tested by selected operators from October 1984 to March 1988 in Berlin city traffic.

Parallel to the development of vehicles, a filling station with storage system and dispenser equipment had to be constructed.

But before testing, plenty of research and development had to be done.

1.2 Basic Research in the Start-Up Phase

In the early beginning of hydrogen car development, a decision had to be made for the H_2 storage system. The classical storage technologies compressed hydrogen with $p_{max} \leq 250$ bar or liquid hydrogen at low temperatures ($T \approx 20K$) were usual and well known. At the beginning of the seventies in the past century, H_2 storage by special metal hydrides were discovered by Reilly and Wiswall [2] especially with iron-titanium. A short time later, some other metal hydride systems were found, which can store hydrogen at ambient temperatures ("low-temperature hydrides"):

- AB systems like FeTi, but with modified alloy compositions;
- AB_2 systems like $TiMn_2$ and modified alloy compositions;
- AB_5 sytems like $LaNi_5$ or $CaNi_5$ and modifications.

In all cases of these metal hydrides, the hydrogen is chemically bonded within the crystal lattice. The reaction is exothermic when the hydride phase is formed, and endothermic when hydrogen is released:

$$\text{metal} + H_2 \rightleftharpoons \text{metal hydride} + \text{heat} \begin{array}{l}\rightarrow \text{released} \\ \leftarrow \text{input}\end{array}$$

The kinetics of this reaction is—in the case of low-temperature hydrides—determined by the heat exchange and not by the steps of adsorption, dissociation, absorption and diffusion of hydrogen. Therefore, a sufficient heat exchange has to be realized for a technical hydrogen storage system.

For the use in vehicles, the exhaust gas and the cooling water are available in principle as heat sources. The charging and discharging pressure during hydrogen exchange is independent of charging condition at constant temperature in the ideal case, and according to the law of mass action it is described by the equation:

$$\ln p = -\frac{\Delta H}{RT} + \frac{\Delta S}{R}$$

ΔH is the binding enthalpy of hydrogen in the hydride and ΔS is the entropy difference between gaseous and bound hydrogen. It is nearly the same for practically all metal hydrides ($\Delta S \approx 120/\mathrm{mol\ K}$). T is the absolute temperature and R is the universal gas constant.

Due to the chemical bonding of hydrogen in the hydride material, hydrogen can only be released by input of heat. This physical property has an additional positive effect for the safety of a metal hydride storage system. For this reason, Daimler-Benz decided to use metal hydride storage systems and to improve the storage capabilities by further basic research and development.

1.2.1 Basic Research of Metal Hydrides

For vehicle applications, a hydride material must permit a cold start and a heat exchange by normal coolants. Depending on the composition, the operating temperature for low-temperature metal hydrides is between −30°C and +80°C with a binding enthalpy of approximately 25 to 30 kJ/mol H_2.

In the beginning of the development of low-temperature metal hydrides at Daimler-Benz, the main interest was concentrated on hexagonal AB_2 systems (Laves Phases) since cubic AB systems (e.g., TiFe) have two different pressure plateaus at low and high H_2 concentrations (Mono-hydride and Di-hydride). This behavior involves problems of cold starting at low concentrations and heat exchange problems for the heating process at high concentrations.

AB_5 systems (e.g., $LaNi_5$) are not suitable for mobile applications due to their high unloaded weight.

With respect to the intended application, the Laves Phases hydrides had to be optimized concerning the following parameters:

- Storage capacity,
- Kinetics of H_2 exchange,
- Sensitivity to impurity gases,
- Multi-cycling stability,
- Activation procedure (incl. re-activation after poisoning),
- Production possibility on the industrial scale (with reasonable price).

After carefully investigating the basic relationships between electron structure and lattice distances (alternation by varying the alloy composition) of Laves Phases and of 3D transitions metals on the one hand, and hydrogen storage concentration and hydriding behavior on the other hand, an alloy consisting of titanium, zirconium, vanadium, iron, chromium, and manganese (exact composition: $Ti_{0.98}$ $Zr_{0.02}$ $V_{0.43}$ $Fe_{0.09}$ $Cr_{0.05}$ $Mn_{1.5}$) was selected to be the most suitable for the application in motor cars.

The basic considerations including detailed measurements of kinetics, diffusion constants, and sensitivity against poisoning gases are reported by Bernauer et al. [3].

The pressure-concentration isotherms (Figure 1.1) of a mass-produced alloy show that a reversible H_2 capacity of 1.8 wt.% related to the initial metal weight can be expected where the charging pressure is 50 bar at a hydride temperature of 24°C. The system can be discharged against a pressure of 1 bar with a hydride temperature above 48°C. The hydride formation enthalpy increases continuously from 22 kJ/mol at low concentrations to 29 kJ/mol at high concentrations. Kinetic measurements of the H_2 reaction show that the dynamics of the system as a whole depends on the heat exchange within the storage unit (see down).

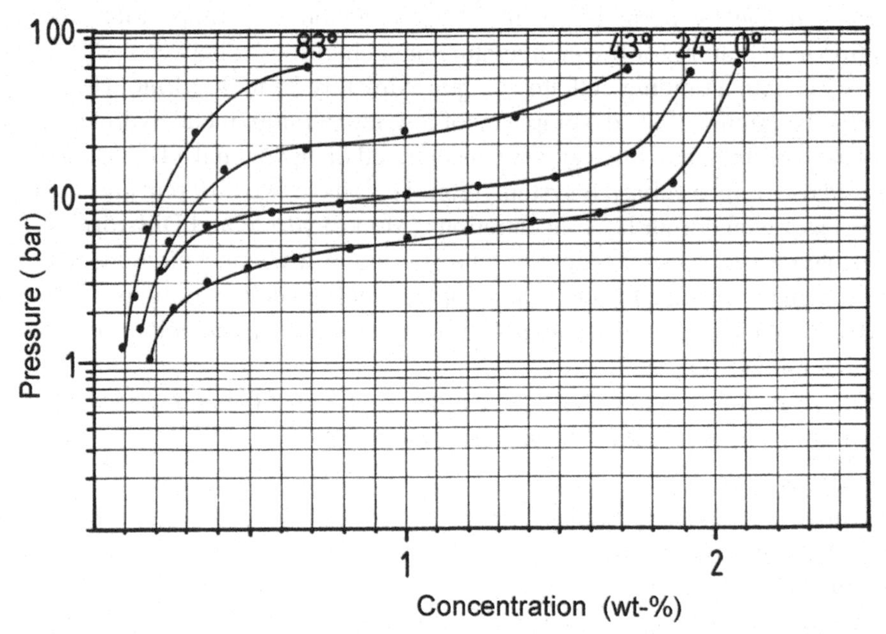

Figure 1.1 Pressure-concentration isotherm of $Ti_{0.98}$ $Zr_{0.02}$ $V_{0.43}$ $Fe_{0.09}$ $Cr_{0.05}$ $Mn_{1.5}$.

Besides the thermodynamic properties, additionally investigated parameters included mainly the stability with respect to extended cycling, the sensitivity to gaseous impurities, and the kinetics of hydrogen exchange. The details of these investigations are described in Reference [4]. But the general result was that the hydride material described earlier fulfilled all requirements for the further technical use in hydrogen vehicles.

1.2.2 Development of the Metal Hydride Container

For the construction of the metal hydride container, some additional physical topics or problems need to be considered:

- During the hydrogen uptake, the crystal lattice of metal expands so that the volume increases. The shape of the vessel must not be deformed by this process.
- Due to this expansion, the generally very hard material embrittles and decomposes to powder with a particle size of $\approx$1 μm.
- Fine powders have a low heat conductivity. Therefore, the velocity of hydrogen exchange becomes slow, because the kinetics of hydride formation and decomposition depends on the velocity of heat transfer.
- Before the first operation, a metal hydride system has to be "activated." By this procedure, the surface layer of poisoning gases adsorbed on the metalsurface has to be removed by heating and evacuation.

In order to meet these requirements all together, a number of different constructions and procedures were tested. This was published in detail in References [5, 6]. In this chapter, only the final solution can be described:

The shape of the final basic container was a cylindrical tube (see Figure 1.2).

For automotive application the diameter is $\phi = 50$ mm. (50×1.2 mm, DIN 2462)

The storage material was milled to a size of $\approx$100 μm, mixed with 5% aluminium powder (for improving heat transfer) and compacted to a stable cylindrical pallet by high pressure. Additionally, this pallet had a concentric cylindrical hole and another free volume that is filled later, when during the activation process the metal hydride expands due to the absorption of hydrogen. The cylindrical hole is for the filter tube made of sintered steel with a porosity of <1 μm (to avoid penetration of hydride particles). In the final storage, tube between two pallets, an additional aluminium cassette (consisting of a cylindrical top and a bottom part) is installed in such a way that it fits closely at the inner surface of the storage tubes and just as closely at the inner

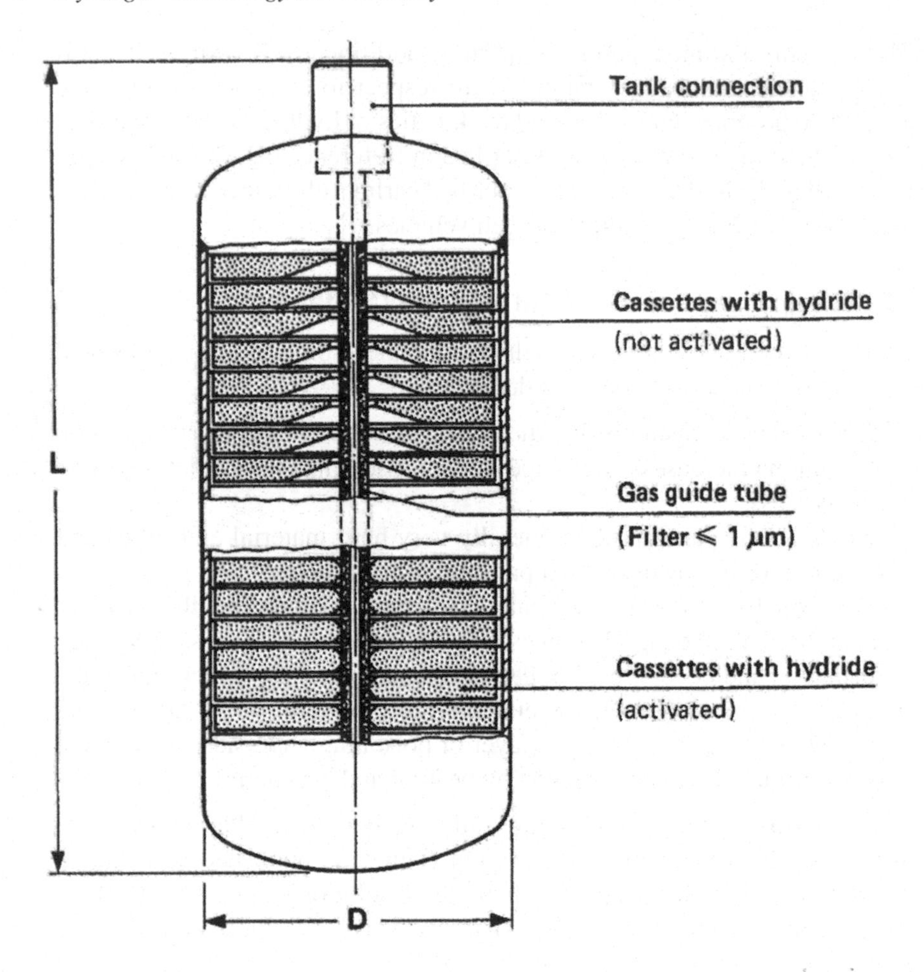

Figure 1.2 Hydride storage unit (schematic view), use of palettes, cassettes, and filter tube.

filter tube. By this construction, these components intensify the cooling and heating effect on the metal hydride in the pallets.

For automotive application, a complete module of hydride storage tank (Figure 1.3) consisted of 19 single tubes like those described earlier.

The individual hydride tubes are interconnected on the gas side by a collection tube.

An outer shell encloses the tube bundle pack in such a way that a heat exchanging medium can pass between the tubes and the outer shell. Filling bodies narrow the hydraulic cross section in order to optimize the heat exchange.

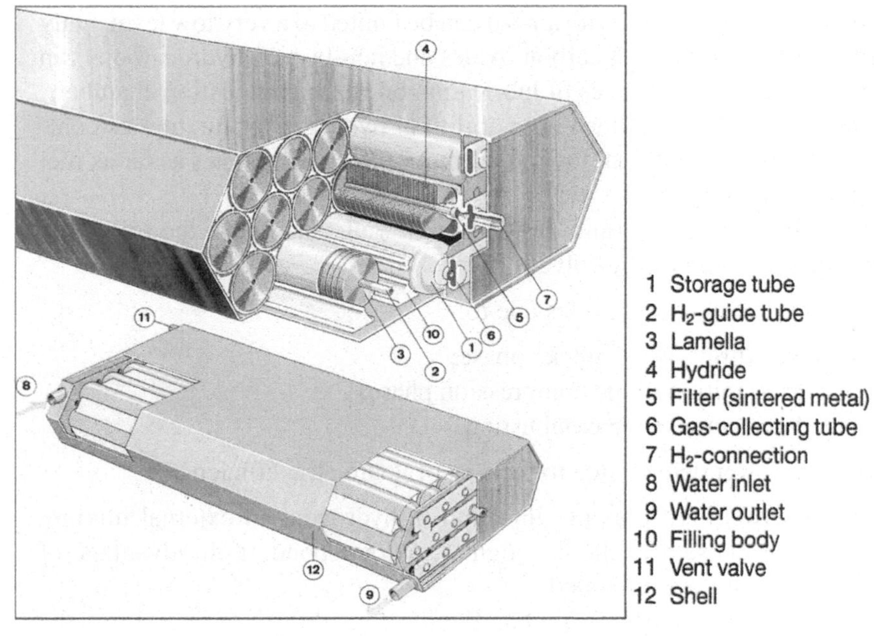

Figure 1.3 Hydride storage tank (module).

The performance data are specified as follows for the use in motor vehicles:
H_2 capacity per module: 16.5 Nm^3 hydrogen

Filling the storage units at the service station:

- Filling time: 10 minutes (80% filling)
- Temp. cooling medium: 18°C
- Quantity cooling water: 0.7 l/s (per module)
- Filling pressure: 50 bar

Discharging the storage units while driving:

- Discharge: 1,600 l/min (max)
- Operating pressure: 2.5–10 bar
- Temp. heating medium: 80°C (max.)
- Quantity heating medium: 0.5 l/s

1.2.3 Development of the H_2 Combustion Engine

Hydrogen is an ideal fuel for internal combustion engines with regard to pollutant emissions. Mainly water vapour is formed. Although the formation of nitrogen oxides cannot be avoided when burning hydrogen—as it is the case for any type of combustion with air—it can be limited to a very low level. Only extremely low quantities of carbon oxides and non-burned hydrocarbons can be formed as a result of traces of lubricating oil in the combustion chamber.

Furthermore, the high lean burn ability of hydrogen/air mixture as compared to that of conventional fuel/air mixtures offers advantages as far as fuel consumption is concerned.

However, there are certain drawbacks as well. They pose the following problems, which have to be solved:

- Irregular combustion cycles due to:
  - Backfiring in the intake phase;
  - Pre-ignition in the compression phase;
  - Knocking during combustion.
- Lower power density due to the loss of volumetric efficiency

In order to avoid high-pressure injection of hydrogen, an external mixture formation was chosen for the H_2-engines. A corresponding disadvantage of low-power density was tolerated.

Without raising the knock limit and/or lowering the pre-ignition limit, the engine can only be operated without any problems within a small speed range. Measures for avoiding irregular combustions can be:

- The addition of ballast through lean-burn operation or exhaust gas recirculation;
- Water injection for cooling the cylinder;
- Cooling the intake air.

As the best solution for the demonstration project, a compromise was chosen with a lean-burn operation in the partial load range and water injection into the suction manifold in the upper load range. This resulted in a compression ratio, which was favourable for both consumption and power, and backfiring was safely avoided.

1.2.4 Development of System Integration

For the demonstration fleet to be operated in Berlin, two different vehicles were constructed, both with metal hydride storage system and internal combustion engine:

- Delivery van on the basis of type 310 with pure hydrogen operation
- Station wagon on the basis of 280 E with hydrogen-gasoline mixture operation

In both cases, a different system integration had to be constructed.

1.2.4.1 System integration for the H_2-fuelled delivery vans with H_2 operation

For hydrogen storage, the delivery vans were equipped with four storage modules as described earlier, two on each side of the vehicle (Figure 1.4). By this storage system, a quantity of 66 Nm^3 of hydrogen was on board of the vehicles (corresponding to an energy equivalent of 22 l gasoline).

In order to heat the storage units while driving and to provide the heat required for hydrogen discharge, heat is removed from the exhaust gas by a heat exchanging system. A recirculating pump transports the heat transfer medium from the heat exchanger to the storage system. The pressure of the storage is controlled by the heat supply to the storage unit.

To fill the storage units, the vehicle is equipped with two connections for coolant supply and return flow, via which the heat of hydride formation can be returned to the filling station for further use.

In addition to hydrogen, the H_2 engine also requires the injection of water (in winter with ethanol as antifreeze) into the intake manifold to prevent backfiring in the higher load range.

1.2.4.2 System integration for the station Wagon 280 E with H_2/gasoline mixture operation

The station wagons were equipped with two H_2-storage modules with a capacity of 33 Nm^3 of hydrogen. The system for heat exchange was similar to the system of the van using the heat for desorption from exhaust gas (Figure 1.5). Additionally, the wagons had a gasoline tank with reduced size.

In the low-load range, the vehicle was operated with hydrogen. With increasing load, more and more gasoline was used while simultaneously reducing hydrogen consumption so that in the case of full load only gasoline was used. Due to this a water injection could be avoided, and on the other hand, the power density of the engine was higher at higher load due to the higher energy density of a gasoline-enriched air-fuel ratio.

1.3 Test Operation of Hydrogen Fleet

In the first step of testing, several prototypes of both hydrogen vehicle types were tested on test benches, in crash tests (for safety purposes) and in real road traffic ($\approx$380,000 km). In this phase, several improvements of construction, design, and handling could be done.

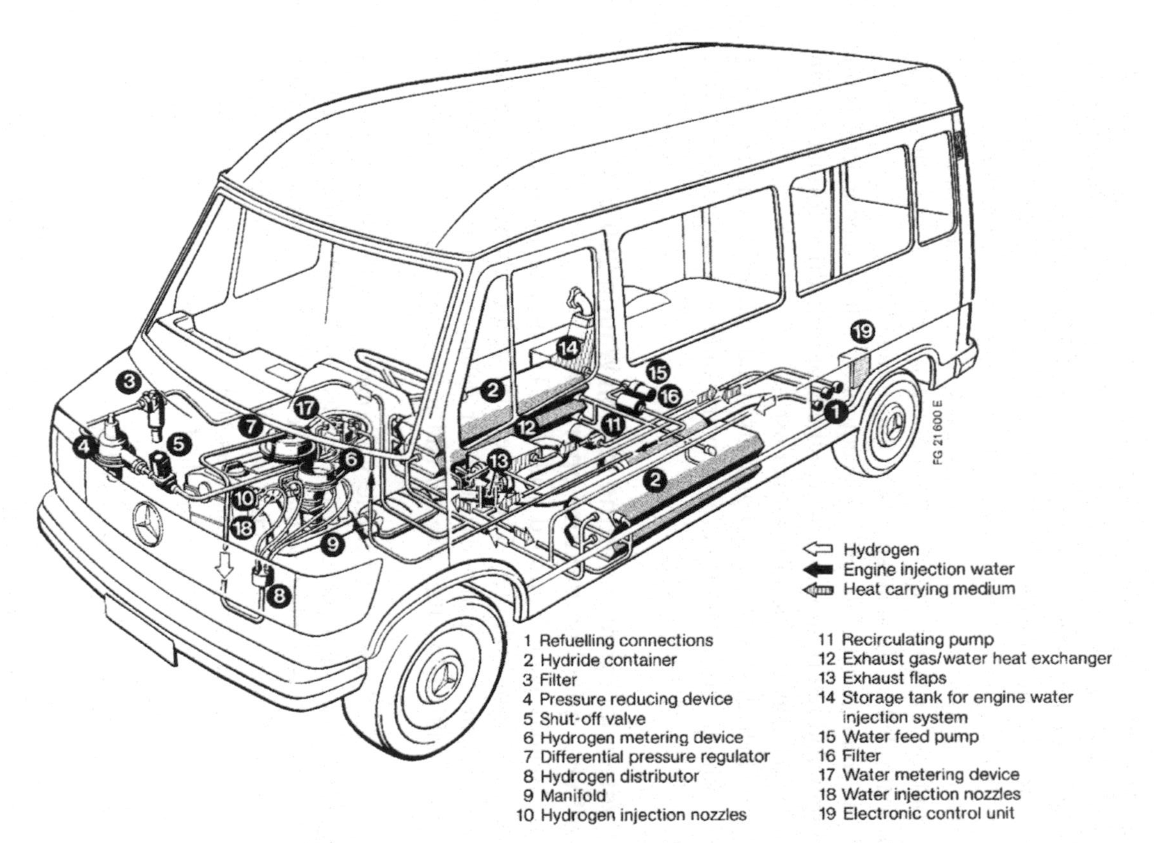

Figure 1.4 Design of the Mercedes-Benz van 310 for H_2 operation.

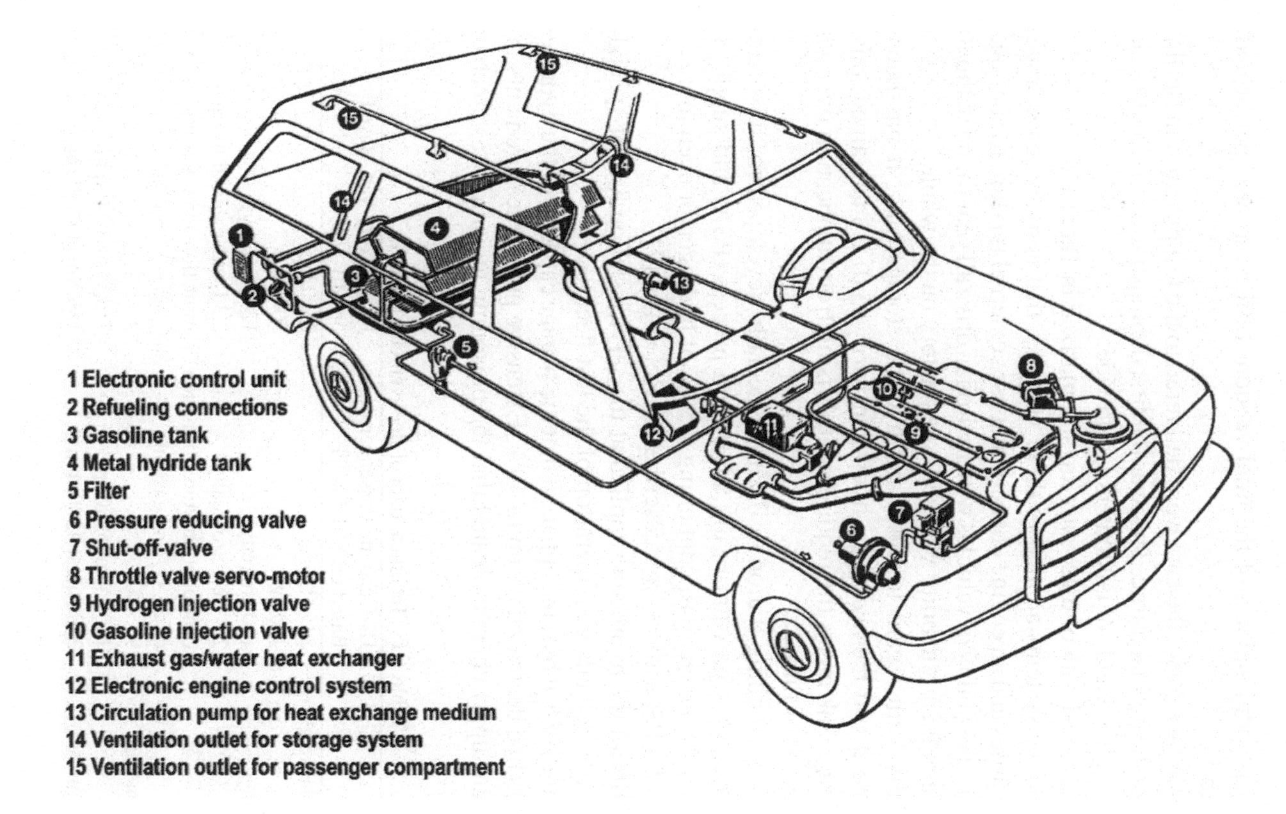

Figure 1.5 Design of Mercedes-Benz 280 TE for mixed hydrogen/gasoline operation.

After final design freeze, five cars of each type were produced for the final use in the hands of customers in Berlin. But before delivering the cars to the customers, an additional test of the final versions of the cars was performed (≈29,500 km).

Starting in Oct. 1984, the cars were delivered to the test program in Berlin and driven in the normal traffic. This was the first demonstration of a hydrogen vehicle fleet in the hands of customers worldwide.

Some technical data of these cars are summarized in Table 1.1.

All technical systems applied in these vehicles, that is hydrogen storage in metal hydrides and an engine concept that uses external mixture formation, confirmed the feasibility of vehicles with hydrogen drive all year round without limitations and their reliability was very satisfactory on the whole.

Especially with regard to the metal hydride storage tank, it has been demonstrated that if the necessary hydrogen purity (5.0) is maintained, this type of storage tank meets the requirements arising from application over an extended period of time.

The hydrogen filling station in Berlin, especially developed for this fleet test, was based on a pressure-swing adsorption process (PSA) and supplied by the former Berlin city gas, containing ≈50% of hydrogen or even more.

Details of a PSA process are described in Reference [7].

The purity of the hydrogen supplied at the service station was of special importance for the fault-free operation of the test fleet.

In autumn 1986, it became apparent that the storage capacity of the hydride storage units had decreased by up to 40%. Extensive purity measurements in the hydrogen supply chain indicated that the poisoning moisture was created

Table 1.1 Technical data of fleet vehicles in Berlin

	H_2/Gasoline Passenger Car, Based on MB 280 TE	H_2-Delivery Van, Based on MB 310
Gross vehicle weight	2350 kg	3500 kg
Pay load	400 kg	700 kg
Distance range	150 km (city)	120 km (city)
Engine	2.8 l electronically controlled mixture formation of air, H_2 and gasoline	2.3 l with external micture formation and water injection
Compression ratio	9	9
Power	120 kW at 5500 rpm	75 kW at 5500 rpm
Maximum torque	230 Nm at 4500 rpm	156 Nm at 4500 rpm
Storage system	2 Hydride containers 280 kg (∼11 l gasoline) 35 l gasoline	4 Hydride containers 560 kg (–22 l gasoline)

between the outlet of the PSA device and the inlet to the intermediate storage vessel prior to the hydrogen transfer to the filling station—that is, in the high-compression equipment of the supply chain. It was necessary to repair or to replace the compressor unit and to reactivate all hydride storage vessels in the vehicles. In the time needed for this work (from October 1985 until September 1986), only a reduced testing program in Stuttgart was possible. (Figure 1.6)

In the whole test period (Oct. 1984 to April 88), all fleet vehicles travelled 256,900 km. In combination with the previous tests of prototypes, about 667,500 km were driven.

More details about the complete test program was published in an explicit report [8].

The program was supported by the German Ministry for Research and Technology.

The goal of the program to investigate the capability for using hydrogen vehicles by normal operators in daily traffic for a long period of time in all seasons was fully reached and the possibility of a safe use of hydrogen in mobile application was demonstrated.

Only very small failures of components occurred and they could be repaired with limited efforts. The damaged compressor in the hydrogen supply chain could be replaced.

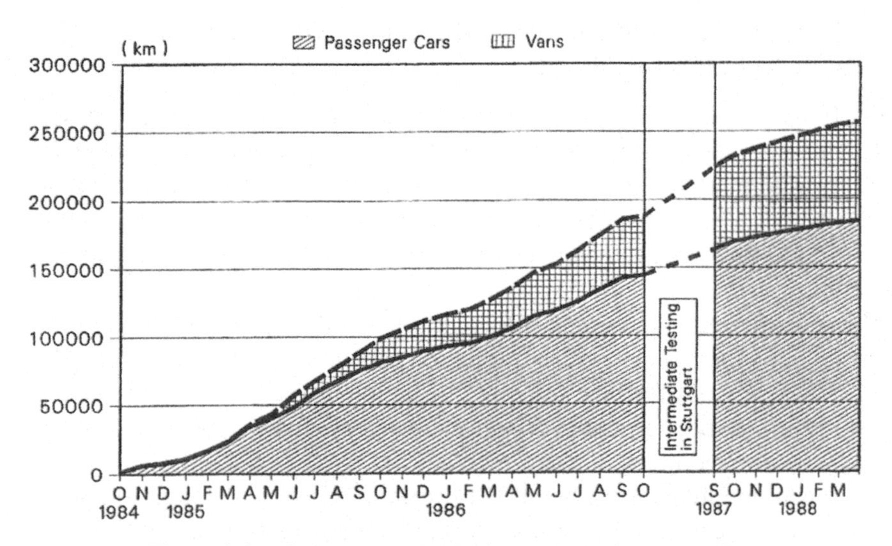

Figure 1.6 Distance travelled by fleet vehicles in road traffic.

The partially deactivated storage vessels due to the poisoning moisture could be reactivated by evacuation and heating.

After the planned end of the program in 1988, the problems of the energy crises were believed to be solved by the responsible decision makers in policy and industry. Therefore, a number of programs for the investigation of alternative propulsion systems were not prolonged, including this project on hydrogen vehicles.

But nevertheless, the general ability of hydrogen as a clean and safe alternative fuel for mobile applications was demonstrated, as well as the availability of its technical and commercial realization.

1.4 Further Developments of Hydride Applications

In addition to the use of metal hydrides in automotive systems, further applications were investigated.

In the first study, which was initiated and funded by the EU and titled "Hydride Storage Devices for Load Levelling in Electric Power Systems," the general capability and the technical conditions for the use of hydrogen in combination with storage in hydrides for huge power plants were investigated up to a H_2 capacity of 320,000 Nm^3. The detailed results were published in Reference [9].

1.4.1 Construction and Test of a Stationary Storage System with 10 t Metal-Hydride

On the basis of this study, a hydrogen storage system for a H_2 capacity of 2,000 Nm^3 was constructed and built (1983–85)—again funded by the EU. As the hydride material, the same composition was chosen as for the motor vehicles ($Ti_{0.98}\ Zr_{0.02}\ V_{0.43}\ Fe_{0.09}\ Cr_{0.05}\ Mn_{1.5}$), because this material was found to have the best characteristics for the required pressure–temperature conditions.

The construction of the single storage tubes was the same as in the case of storage systems for vehicles with compressed pallets of metal hydride powder (incl. 5% Al- powder for heat transfer). Additional cassettes of aluminium improved the heat transfer, and hydrogen exchange was carried out by a central filter tube of sintered metal.

Seven storage tubes were bundled to form a cylindrical storage module (Figure 1.7) in a construction similar to the automotive storage with filling bodies for optimising the heat exchange.

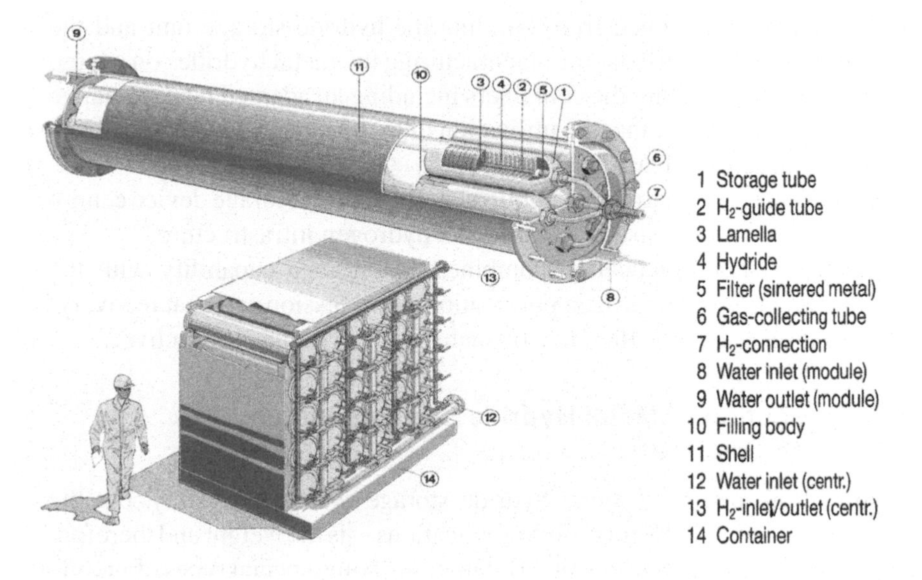

Figure 1.7 Metal hydride storage tank for stationary applications (model with a 5 × 6 array of modules).

The final storage system was constructed with 32 modules in 8 vertical line-ups with 4 modules each. Details of construction are too manifold to be reported in this short overview, but all data and specifications of manufacturing and operation (in 1985–87) are published in Reference [10], including dimensions, pipework for hydrogen and heat exchange medium, valves, safety components, etc.

In a short overview, the results of construction and demonstration of this large-scale hydride storage system can be summarized:

- Metal hydride technology can usefully complement the existing systems of hydrogen supply and utilization, for example, in the hydrogen pressure range 40–200 bar, even in this large scale.
- One of the outstanding features of hydride storage devices is their ability to absorb hydrogen to a high (volumetric) storage density, at low feed pressures, and at ambient temperatures. The pressure plateau of a metal hydride can be adapted to individual application and safety requirements.
- Hydride storage systems comply in principle with existing specifications and technical regulations concerning pressure vessels and do not represent any additional problems with respect to safety or other technical issues.

- The knowledge gained from building the hydride storage unit and the possibilities, which exist for manufacturing the metal hydrides on a large scale and for building the containers including an adapted heat exchange equipment, indicate that the utilization of hydride storage facilities on an industrial scale is feasible.
- On the basis of the storage function alone, a hydride storage device cannot compete with the existing compressed hydrogen infrastructure.
- However, if an application combines this storage capability with the other functions of hydrogen purification, compression and heat recovery/utilization, the cost-efficiency of such systems becomes attractive.

1.4.2 Application of Metal Hydride Storage System in a Submarine

The special advantage of metal hydride storage systems is safety, but the disadvantage—especially for mobile applications—is the weight and therefore the low gravimetric storage density. But there are some special types of mobile use of hydrogen where the high weight of hydrides does not matter or it is even desired—like in forklifts or submarines. Parallel to the hydride research at Daimler-Benz in Germany as described in this chapter, the application of hydrides in a new concept of submarines was considered at Howaldtswerke Deutsche Werft AG–HDW (today "thyssenkrupp Marine Systems").

In this new concept, the air-independent propulsion (AIP) was designed and realized by fuel cells with a metal hydride storage system, a liquid oxygen container and—until the 1980's—an alkaline fuel cell. In later constructions, PEM fuel cells were used.

With this propulsion system, much longer ranges for submerged operation were reached in comparison to conventional battery systems. An additional advantage of metal hydrides is the need of heat for the desorption of hydrogen. In combination with the waste heat of the fuel cell, the thermal output of the whole system is very low and therefore the thermal signature of the whole boat is extremely low.

As hydride material a similar composition as for automotive applications was selected and even the construction of heat transfer components within the storage tubes was similar.

A diagram of the energy supply system of the boat is shown in Figure 1.8.

This integrated concept will lead to optimum performance for submarines of any size.

The position of metal hydride tanks in the boat outside of the hull is shown in Figure 1.9.

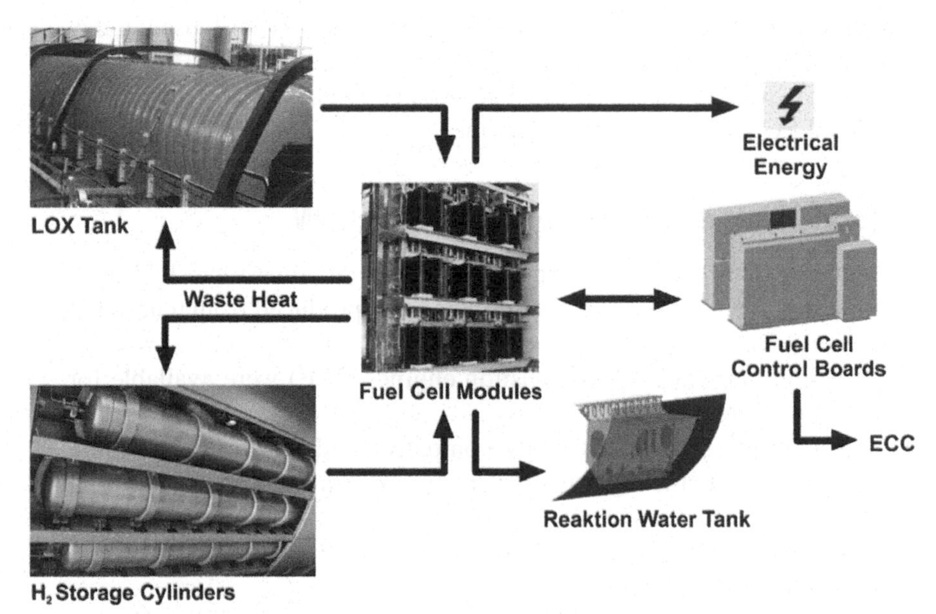

Figure 1.8 A diagram of energy supply system in a submarine with metal hydrides.

Source: © thyssenkrupp Marine Systems.

Figure 1.9 Position of hydride containers in the submarine.

Source: © thyssenkrupp Marine Systems.

The application of metal hydrides in submarines is still used today at thyssenkrupp Marine Systems for different classes of boats. Even the conversion of existing diesel-electric submarines is possible by integrating an additional FC section into the submarine.

1.5 Mobile Application of Hydrogen at BMW

So early in the 80's of the past century, BMW decided to investigate hydrogen as an alternative and sustainable fuel (provided it is produced from renewable primary energies).

At this time, only internal combustion engines (ICE) were available for the propulsion system.

For the hydrogen storage system, the company decided to work with liquid hydrogen (LH_2) due to the high energy density and as a consequence the long range of cars.

Additionally, the development of LH_2 technology had some benefits from the space applications in the USA.

An overview on the whole history of BMW activities with hydrogen vehicles is given in Figure 1.10.

The first chief steps in the early beginning were

- Adaption of a conventional ICE engine to hydrogen fuel with external mixture formation and additional means to avoid backfiring.
- Conception of a LH_2 container system with vacuum insulation valves, regulation, refuelling, and measuring systems.
- Systems integration of all components.
- Sensors and safety systems.

The construction of the tank was an intensive cooperation with the German Aerospace Center, DLR (formerly "Deutsche Forschungs- und Versuchsanstalt für Luft- und Raumfahrt" DFVLR).

A detailed description of the tanks is given by Peschka [11].

The first experimental car was a BMW 520 with a duel fuel supply (H_2 or gasoline), external mixture formation, and turbo charger.

In further developments a number of improvements could be realized:

- Increase of power density of the combustion engine.
- Reduction of NO_x emissions by lean mixture.
- Additional reduction of backfiring tendency by sequential injection.
- Reduction of evaporation rate by improving the thermal insulation of the tank.

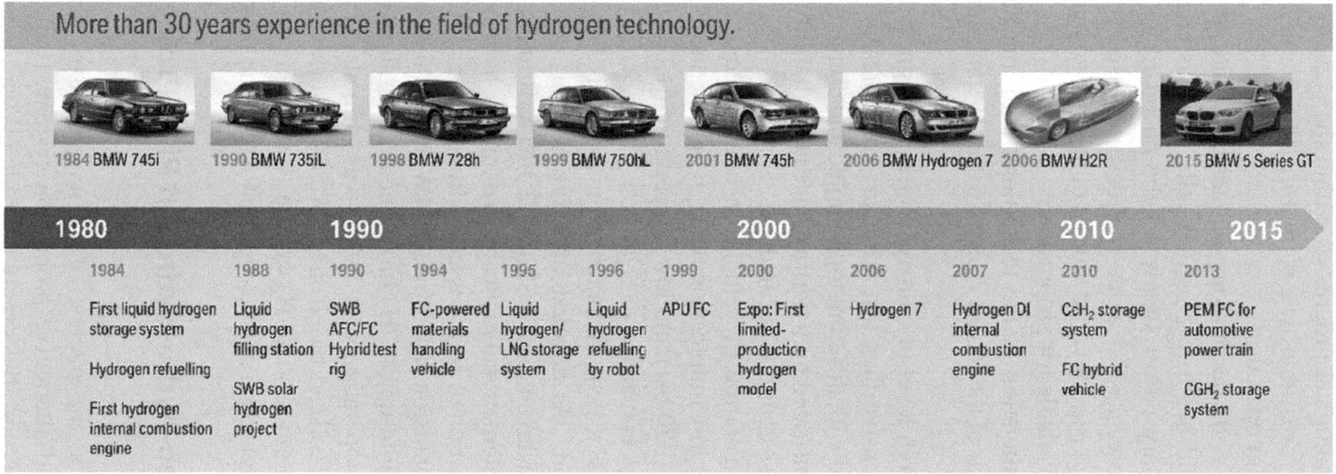

Figure 1.10 Historical milestones of H_2 vehicles at BMW.

Source: © BMW.

In addition to these technological progresses within the cars, the refuelling concept (handling, safety) was improved so that in 1996 the first automatic H_2 filling station could start operation.

At the end of the 90's, the technology of fuel cells had progressed to the point that BMW could begin testing fuel cells in a passenger car—in the first step as an auxiliary power supply for the on board generation of electricity. For this purpose, a solid oxide fuel cell was used for heat and power cogeneration. The generator could thus be eliminated so that the saved power was additionally available for propulsion.

As a consequent continuation of the technology path, the next step of development was a new car of the model 7 series with LH_2 tank and ICE with direct H_2 injection.

On the basis of this high level of construction and design, a fleet of 100 cars was built for an extensive fleet test. The cars were presented in many locations around the world with different topographic and climate conditions. In all cases the cars demonstrated high performance and reliability so that the series-production readiness could be verified.

Parallel to this demonstration of vehicles, a new concept of hydrogen storage was developed: Cryo-Compression (CcH_2). In this concept, hydrogen is stored at high pressure (350 bar) and low temperature (−240°C) so that a high density of up to 72 g/l is reached.

This technology was tested in practical tests. It is approved and available for application. A detailed description can be found in: "http://www.storhy.net/finalevent/pdf/WS3_CcH2_BMW-Brunner.pdf".

In the past years, the international mainstream of mobile hydrogen focused more and more in the direction of compressed hydrogen storage (CH_2) and to fuel cells with electric propulsion. Therefore, BMW joined this technological path to cooperate in the build-up of CH_2 infrastructure. For the development of fuel cells, a strategic cooperation with Toyota was formed.

First prototypes of fuel cell cars based on BMW 5GT (hybridized by battery for recovery of braking energy) and a CH_2 storage system had very successful tests so that a market introduction can be expected within the next years.

1.6 The Euro-Québec-Hydro-Hydrogen Pilot-Project (EQHHPP)

At the end of the 1980's, the interest of German industry with respect to renewable energies and alternative propulsion systems (incl. hydrogen) decreased significantly, because the supply of crude oil was continuous and safe and the emission problems of exhaust gas were intermediately solved by catalysts.

For this reason, no further development or demonstration projects for hydrogen technologies were started in this time frame.

But in the political scene, there were some considerations about the future energy supply with respect to the limited fossil energy sources as a matter of principle.

As one hydrogen-related activity, an intensive study about the intercontinental contribution of hydropower for the energy supply of Europe was launched by the European Commission in cooperation with the Government of Québec and several Canadian and German companies and institutions.

In this "Euro-Québec-Hydro-Hydrogen Pilot-Project" (EQHHPP), it should be investigated how and under which conditions renewable primary energy in the form of hydropower from Québec can be converted via electrolysis into hydrogen. This hydrogen should be shipped to Europe to be stored and used in different ways: electricity/heat co-generation, vehicle, aviation, ship propulsion, and steel fabrication. In a first planning concept, a system with a power of 100 MW was theoretically investigated. Original planning is intended for each step of the project to be carried out: hydrogen production and storage, transoceanic hydrogen transportation, use of hydrogen in a bus fleet, cogeneration plants, and an airbus turbine.

This project was the first investigation of an intercontinental chain of a sustainable energy system with renewable primary energies, intercontinental transportation of energy via a secondary energy carrier with high-energy density and the utilization by the end-user with high efficiency by fuel cells.

The concept of this energy chain is shown in Figure 1.11. The detailed results are described in Reference [12].

As transport medium liquid hydrogen (LH_2) was investigated as well as the methylene–cyclohexane–toluol–system, Toluol can be hydrided (with assistance of a catalyst) to methylene cyclohexane (MCH) and after transportation it can be decomposed again to toluol and hydrogen.

Even if the MCH is no longer being considered for transportation of hydrogen today, the LH_2 path is still current.

In further steps of this project, all details and components of this energy vector were investigated and calculated with respect to efficiencies, cost, stabilities, etc.

The results of these studies are reported in Reference [13].

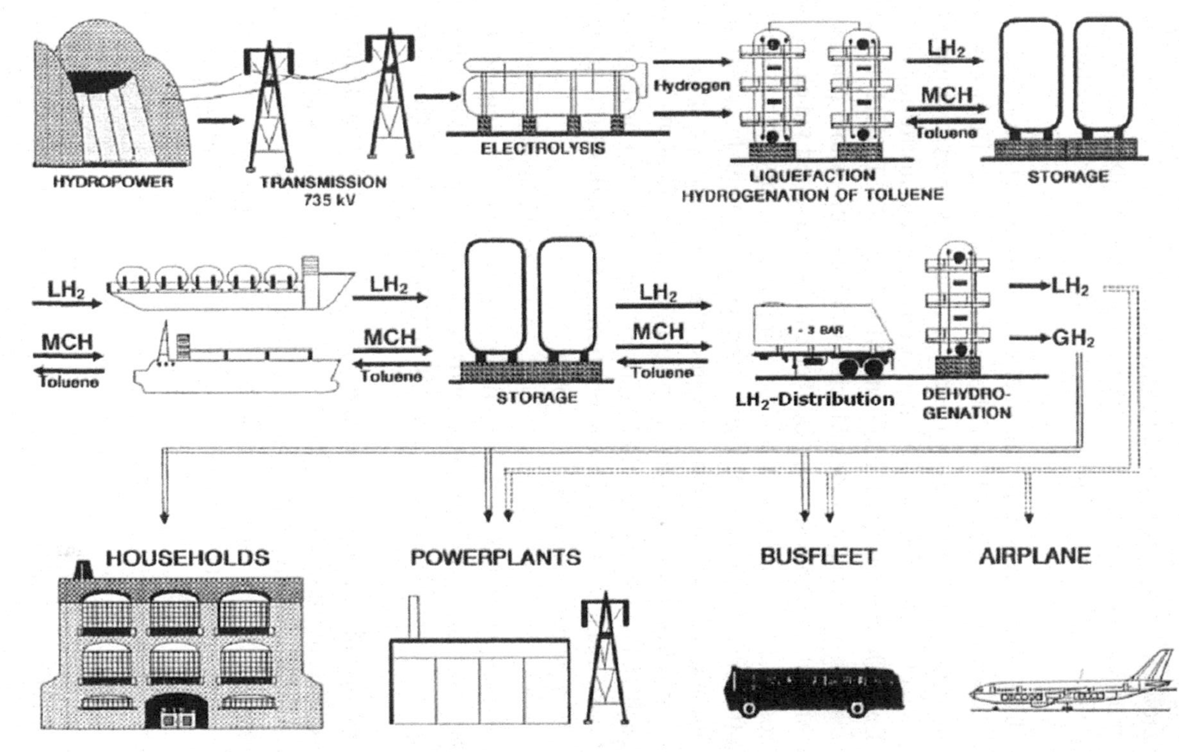

Figure 1.11 The concept of EQHHPP-project.

Source: Euro-Québec Hydro-Hydrogen Pilot Project, Brochure, Commission of the European Communities and Government of Québec, 1989.

The realization of the studies could not be financed at the end, but the results of these investigations are still available and can be accessed in the future if a realization will be necessary for the development of sustainable energy systems.

1.7 Development of Hydrogen Technology with PEM Fuel Cells for Mobile Applications

In the beginning of the 1990's, the further development of hydrogen technology in Germany was quite poor. The problems of the former energy crisis were politically solved and the exhaust gas emissions of fossil fuels could be sufficiently limited by catalysts. In this phase, the industry did not see urgent need to look for new or sustainable energy sources or alternative fuels.

But in the mid of the 1990's, the awareness of limited resources of fossil fuels increased again and the complex of CO_2 emissions and climate change came into the public discussion. Additionally, a new technical approach to solve these problems came into consideration: the Polymer Electrolyte Membrane Fuel Cell (PEMFC) and former types of fuel cells (i.e., Alkaline FC) did not have enough power density for mobile applications.

In this situation a reactivation of hydrogen technology started—mainly promoted by automotive industry.

1.7.1 H_2/FC Technology in Passenger Cars

Most of the vehicle manufacturers in Germany investigated intensively hydrogen propulsion system for passenger cars with different H_2 storage systems and powertrains:

- BMW: Liquid hydrogen (LH_2) and internal combustion engine (ICE) until 2012, later compressed hydrogen (CH_2), PEMFC, and electric engine;
- Opel: LH_2 and CH_2 storage with PEMFC and electric engine;
- Ford: CH_2 storage with PEMFC and electric engine as well as with ICE;
- Daimler: Most intensive research and development with different types of cars, (passenger cars, vans, buses) and different storage systems, all with PEMFC and electric engine (see Figure 1.12). Therefore, the following description of the history of hydrogen mobility in Germany is mainly focused on the Daimler activities.

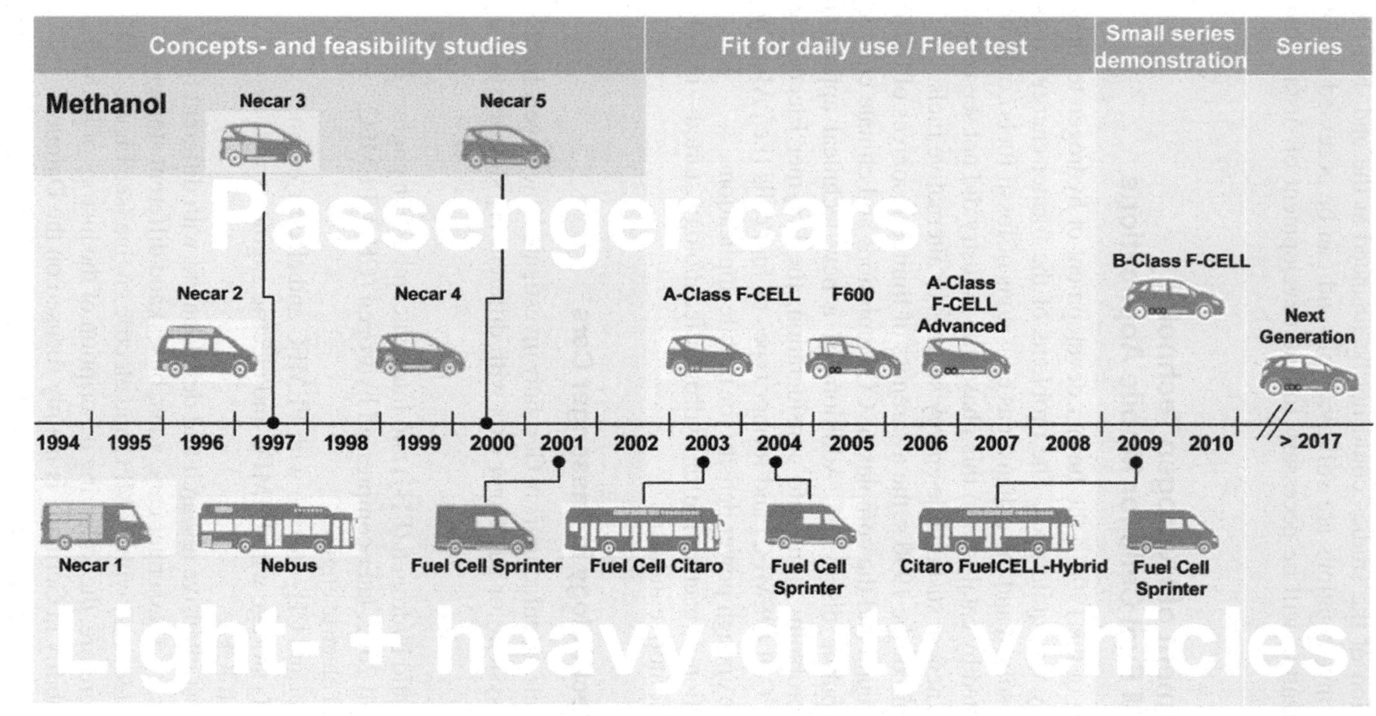

Figure 1.12 History of Daimler fuel cell vehicles since 1994.

Source: Daimler AG ("Necar" = New electric car).

In the first experimental phase (∼1994–2000), compressed hydrogen (350 bar) was used as the storage system in Necar 1 and 2. In Necar 4 it was liquid hydrogen, and in Necar 3 and 5 hydrogen was produced on board by a methanol reformer.

After this testing phase, further developments concentrated on CH_2 storage systems with 700 bar for passenger cars (starting with B-Class) and 350 bar for buses and vans. The storage tanks are completely wrapped, carbon-fiber-reinforced composite vessels with aluminum liner ("type 3 tank") or plastic liner ("type 4 tank"), depending on the supplier. Details of the tank system with valves, sensors, pressure regulator etc. are described by Mohrdieck et al. in Reference [15].

Parallel to the technical development of hydrogen cars and components, the rules for safety and regulation for approval had to be generated.

With respect to safety, the high-pressure components implicate the highest challenges. European solutions for the definition of safety standards were requested in a very early stage. As a first step, the safety requirements and testing procedures were defined by the "European Integrated Hydrogen Project" (EIHP). For the proof of safety of CH_2 containers and system components, a lot of testing and approval conditions have been determined. After several years of experience, the new EU directives [16, 17] were elaborated by "EC_DG_ENTER_H2-V-Reg working group" and decided by the EU parliament and council in 2009/10.

In the years from about 2000 until today, at Daimler, a continuous improvement of different types of vehicles (see Figure A) with respect to range, consumption, efficiency, safety, performance, etc. of components and systems has been achieved. One example of progress for the step from A- to B-Class in 2009 is shown in Figure 1.13.

Additionally, it is remarkable that the new B-Class vehicle is constructed, fabricated, and approved under series production conditions.

In 2011, the performance and reliability of these vehicles were demonstrated by a tour around the world of 3 cars, each with more than 30,000 km in 125 days, with no failure of the propulsion system or any of its components.

At present, a series production of fuel cell cars is in preparation. The presentation of the first series-produced fuel cell passenger cars from Daimler is announced for 2017 on the basis of a GLC-Class.

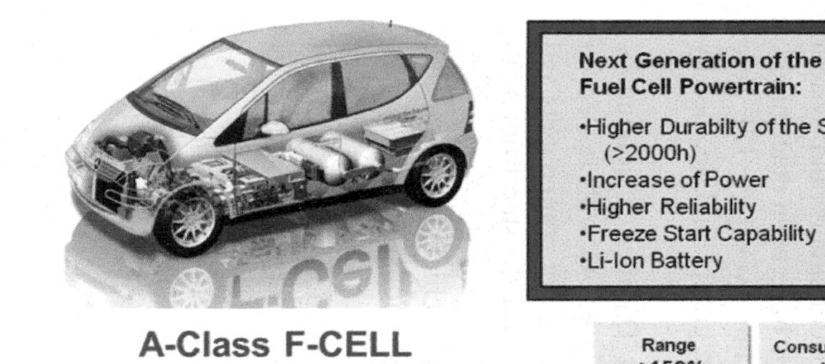

Technical Data	
Vehicle Type	Mercedes-Benz A-Class
FC-System	PEM, 72 kW (97 PS)
E-Motor	Asynchronous Motor Power (cont. / peak): 45 kW / 65 kW (87 HP) Peak Torque: 210 Nm
Fuel	Hydrogen (350 bar)
Range	170 km (NEDC)
V max	140 km/h
Battery	NiMh, Power (cont. / peak): 15 kW / 20 kW (27 PS); Capacity: 6 Ah, 1.2 kWh

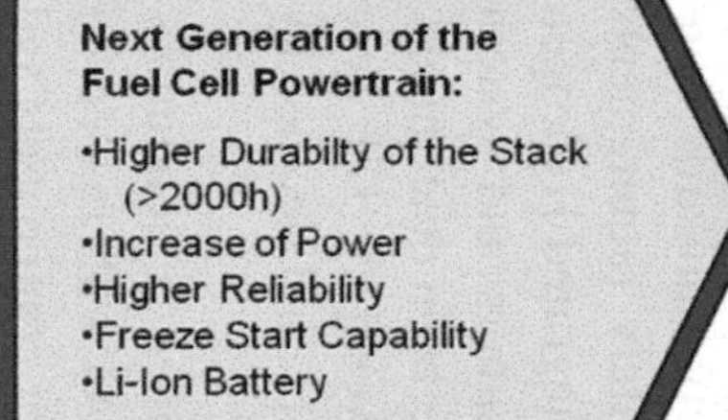

Technical Data	
Vehicle Type	Mercedes-Benz B-Class
FC-System	PEM, 80 kW (109 PS)
E-Motor	Permanent Synchronous Motor Power (cont. / peak): 70 kW / 100 kW (136 HP) Peak Torque: 320 Nm
Fuel	Hydrogen (700 bar)
Range	385 km (NEDC)
V max	170 km/h
Battery	Li-Ion, Power (cont. / peak): 24 kW / 30 kW (40 PS) Capacity: 6.8 Ah, 1.4 kWh

Figure 1.13 Technology enhancement from Mercedes A-Class (2007) to B-Class (2009).

Source: Daimler AG.

1.7.2 H_2/FC Technology in Urban Buses

The H_2/FC technology is very suitable for application in urban buses. These buses typically have a clearly defined driving cycle and come back to their city depot. For cities, the usage of fuel cell city buses offers a possibility to reduce significantly environmental and noise pollution.

The first bus at Daimler was the Nebus (New Electric Bus) in 1997. It was a single experimental vehicle for testing purposes with a

- CH_2 storage system, 300 bar, 7 vessels on the bus roof, capacity 21 kg H_2;
- Fuel cell system with 250 kW (from Ballard);
- Electric propulsion with two integrated well hub motors with 75 kW each.

After the testing period, a first fleet of 36 buses was built for the "CUTE-Project" [18].

Within this project, three buses were operated in daily use in ten Europeancities, in Perth (Australia), and Beijing—in each case in very different topographic areas. The technical data of these buses are summarized in Figure 1.14.

All buses together made more than 2.1 Miokm in ca. 139,000 hours of operation with high performance.

The next generation of fuel cell buses was the Citaro FuelCell-Hybrid with significant improvements with respect to the predecessors (Figure 1.14). These buses are in daily operation in several cities (e.g., Hamburg).

In the past years in addition to Daimler, other bus manufactures constructed and built hydrogen buses, which are operated in different European cities in daily traffic.

An overview can be seen in "*http://chic-project.eu/fuel-cell-buses-in-europe*".

In this summery not only the technical data of hydrogen buses are presented but also the social aspects of operation, the public acceptance, and the subjective well-being of passengers and drivers.

1.7.3 Clean Energy Partnership

The Clean Energy Partnership (CEP) was established in December 2002 as a joint initiative of government and industry, headed by the German Ministry of Transport and Traffic. Its aim is to test the suitability of hydrogen as a fuel. Since 2008, the CEP has been a lighthouse project of the National Hydrogen

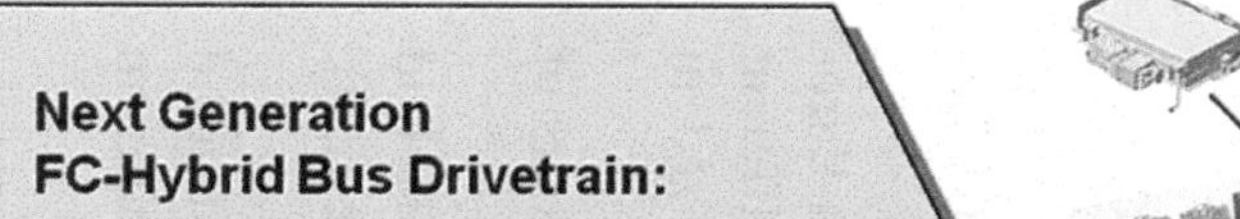

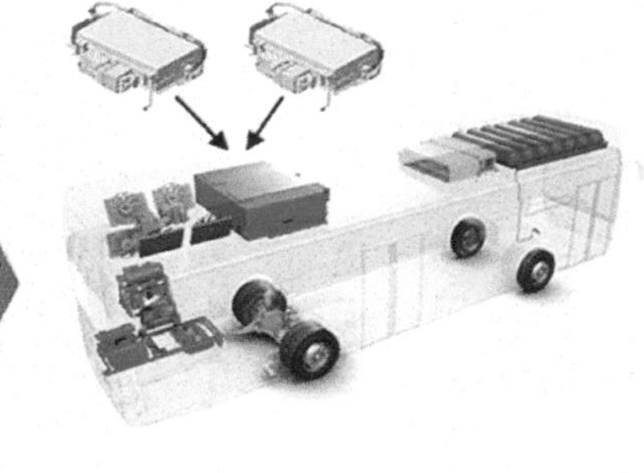

FC-Bus (CUTE)

Technical Data	
Power FC-System	250 kW
Lifetime (FC)	4 Years
Power E-Motor	205 kW, for < 15-20 sec
Hydrogen Tank	40 – 42 kg Hydrogen (350 bar)
Range	180 - 220 km
HV-Battery	--
Efficiency FC-System	43 - 38 %
H2 Fuel Consumption	20 – 24 kg / 100 km

Citaro FuelCELL-Hybrid

Technical Data	
Power FC-System	120 kW (cont.) / 140 kW (max.)
Lifetime (FC)	6 Years
Power E-Motor	Power (cont. / peak): 2 x 80 kW / 2 x 120 kW
Hydrogen Tank	35 kg Hydrogen (350 bar)
Range	> 250 km
HV-Battery	26,9 kWh, Power: 250 kW
Efficiency FC-System	58 - 51 %
H2 Fuel Consumption	10 – 14 kg / 100 km

Figure 1.14 Improvements of the Mercedes-Benz Citaro FuelCELL-Hybrid Buses compared to the predecessor Citaro-Bus of the CUTE-Program.

Source: Daimler AG.

and Fuel Cell Technology Innovation Programme (NIP), implemented by NOW (National Organization for Hydrogen and Fuel Cell Technology, see "*https://www.now-gmbh.de/en*").

The CEP is the biggest demonstration project in Europe in the field of hydrogen technology. Twenty industry partners—Air Liquide, Bohlen & Doyen, BMW, Daimler, EnBW, Ford, GM/Opel, H_2 Mobility, Hamburger Hochbahn, Honda, Hyundai, Linde, OMV, Shell, Siemens, Stuttgarter Straßenbahnen SSB, Total, Toyota, Volkswagen and Westfalen—are involved in the CEP. Experts from these companies work together in councils that transcend branch boundaries to pave the way for the market launch of hydrogen vehicles.

This includes the following fields of activities:

- Production: Clean and sustainable production of hydrogen and storage of liquid and gaseous hydrogen;
- Infrastructure: Defining technical standards for fueling process and stations, Expansion of thc network of filling stations; Quick and safe refueling;
- Mobility: Continuous operation of efficient hydrogen vehicles.

Today, the following vehicles or supply systems are operated within the CEP project:

- 115 passenger cars from BMW, Daimler, Ford, VW, Honda, Hyundai with electric powertrains with fuel cells and CH_2 storage vessels (up to 700 bar);
- 20 buses from Daimler and Solaris with 350 bar CH_2 storage vessels;
- 20 fuelling stations in various cities (see: www.h2station.info);
- The number of filling stations will be expanded to 50.

For the expansion of the infrastructure, the CEP cooperates with some German federal states and the H_2-Mobility company (see Figure 1.15).

H_2-Mobility is a company consisting of car manufacturers, gas companies, and operators of filling stations.

1.7.4 Special Mobile Applications

Besides the main stream of mobile applications with passenger cars and buses, a lot of other mobile H_2/FC propulsion systems (wheelchairs, electro-bikes, fork lifts, material handling vehicles, ...) are developed and tested by different organizations (university institutes or companies). The main goal of this work is to gain more experience with new systems and to make the first steps of market introduction for special niches.

Figure 1.15 CEP & H_2 Mobility—Expansion of the filling station network in Germany. For further details of CEP see www.cleanenergypartnership.de.

Within this chapter two examples can be presented:

- **A**: E-bike with fuel cell and CH_2 tank, constructed and built by students at the "University of Applied Sciences Rhine-Main in Rüsselsheim" in cooperation with some industrial partners. (Figure 1.16)

This bike is still a prototype with the following technical parameters:

Power of fuel cell:	360 W (Schunk Bahn- und Industrietechnik GmbH)
Power of electric engine:	250 W (Gernweit)
Capacity of tank:	0.001 Nm^3 at 300 bar (Armotec S.R.O)
Range	~140 km depending on topography and weight of person

The next steps for commercial use will be a small series production and the assembly of an infrastructure for hydrogen supply.

- **B**: Another development for special application of H_2/FC vehicles is materials handling vehicles for factories, big storage facilities, or magazines. They combine the advantages of zero emissions in halls and high operation ranges in open-air grounds (like airports).

Figure 1.16 E-bike with fuel cell and CH_2 tank.

One first example is a tractor, constructed (incl. systems integration) and operated by Bosch based on a conventional Mulag tractor (see Figure 1.17).

The essential parameters of this tractor are:

- PEM Fuel Cell with power P = 10 kW in range extender operation
- LiTi battery with rated power P = 30 kW for operation and recuperation
- Capacity E = 6 kWh
- H_2 storage capacity $c(H_2) = 3.2$ kg at $p_{max} = 350$ bar
- Filling time for H_2 tank $t_{tank} \leq 4$ min
- Max. time of operation $t_{max} = 8$ h

This prototype vehicle is currently in a test and demonstration phase, sponsored by a publicly funded project "Innovative regenerative on-board energy converter" (InnoRobe).

Another larger project for the development of materials handling vehicles is a cooperation of Austrian and German companies (DB Schenker, Fronius, HyCentA, Joanneum Research, Linde MH, OMV), named "E-Log-Bio-Fleet."

In the first project period "E-Log-Bio-Fleet" in 2011–2013, hydrogen was extracted from biomass and the first vehicles were just tested in daily operation by DB Schenker. This project was prolonged until 2016 to get more real-life experience, especially in terms of the fuel cell lifetime.

Figure 1.17 H_2/FC-tractor.

Source: © Bosch.

Furthermore, a very popular project in materials handling is the so-called "H2IntraDrive" project.

In this project, certified H_2 mainly produced from waste H_2 from chemical processes or from renewable primary energies was used to supply a test fleet of counterbalanced forklift trucks and tow trucks as shown in Figure 1.18.

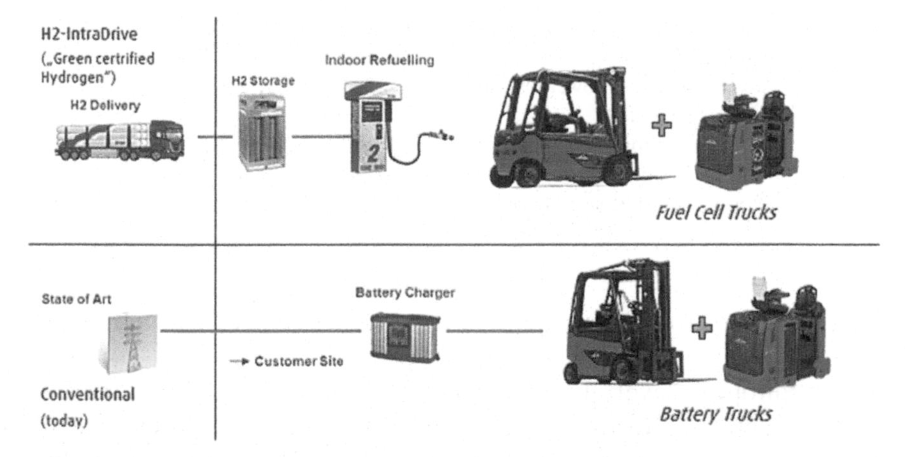

Figure 1.18 Comparison of conventional electric trucks with H_2 trucks of H_2 intradrive-project.

Source: Linde MH.

The first trucks were commissioned in 2014 to test and demonstrate the profitability, reliability, and sustainability in daily use.

This project can be seen as the next step toward market penetration. In a first application in a production facility, five forklifts and six tow-trucks were employed in the BMW factory in Leipzig, see Figure 1.19.

Some technical details of the fuel cell system of forklift trucks are listed in Table 1.2.

Figure 1.19 Forklift and tow-truck (Linde MH) in the BMW factory in Leipzig in practical use.

Source: Linde MH.

Table 1.2 Technical details of forklifts (387-series) from Linde MH

Nominal Voltage	80 V
Max. Continuous Power Output	10 kW (13.41 hp)
Hydrogen Capacity	1.8 kg (3.97 lb)
Hydrogen Nominal Pressure	350 bar (5076 psi)
Refilling Time	90 sec.
Hybrid Energy Storage Type	Lithium Ion Batt.
Required Level of Hydrogen Purity	5.0

The experiences were very positive so that further applications are to be expected.

On the European level, the importance of materials handling application of H_2 mobility is recognized. Several activities are supported under the umbrella of the European "HyLift" projects. In these projects, up to 200 vehicles are planned to be tested or operated in demonstration phases in more than 10 European sites. These activities are described in detail in *www.hylift-projects.eu.*

One main goal of this EU project is to prepare the market penetration of hydrogen technology in this special market where a suitable ratio of vehicle/filling station is possible: with one filling station several vehicles with a long daily time of operation can be supplied so that the H_2 supply can be realized commercially even in the start-up phase.

1.8 Stationary Applications in Households

In Germany, nearly 30% of primary energy consumption is needed by households. This energy demand is allocated to heat and electric power. Up to now, electricity is distributed by the electric grid, and heat is locally obtained from natural gas, oil, or coal.

These different energy paths involve high energy losses, typically shown in an example from Vaillant in the upper part of Figure 1.20.

In a first strategic step, these losses can be reduced by local cogeneration of heat and power. Currently, this could be done by an internal combustion engine (powered mainly with natural gas in households). Due to the Carnot efficiency of the ICE, the electric output is limited to ~30%.

In a second strategic step, the primary efficiency could be further increased by using natural-gas-fueled fuel cells instead of ICE's. This is shown in the lower part of Figure 1.20 with typical values for the efficiencies. There are two possibilities to use natural gas as fuel in fuel cell systems for cogeneration of heat and power.

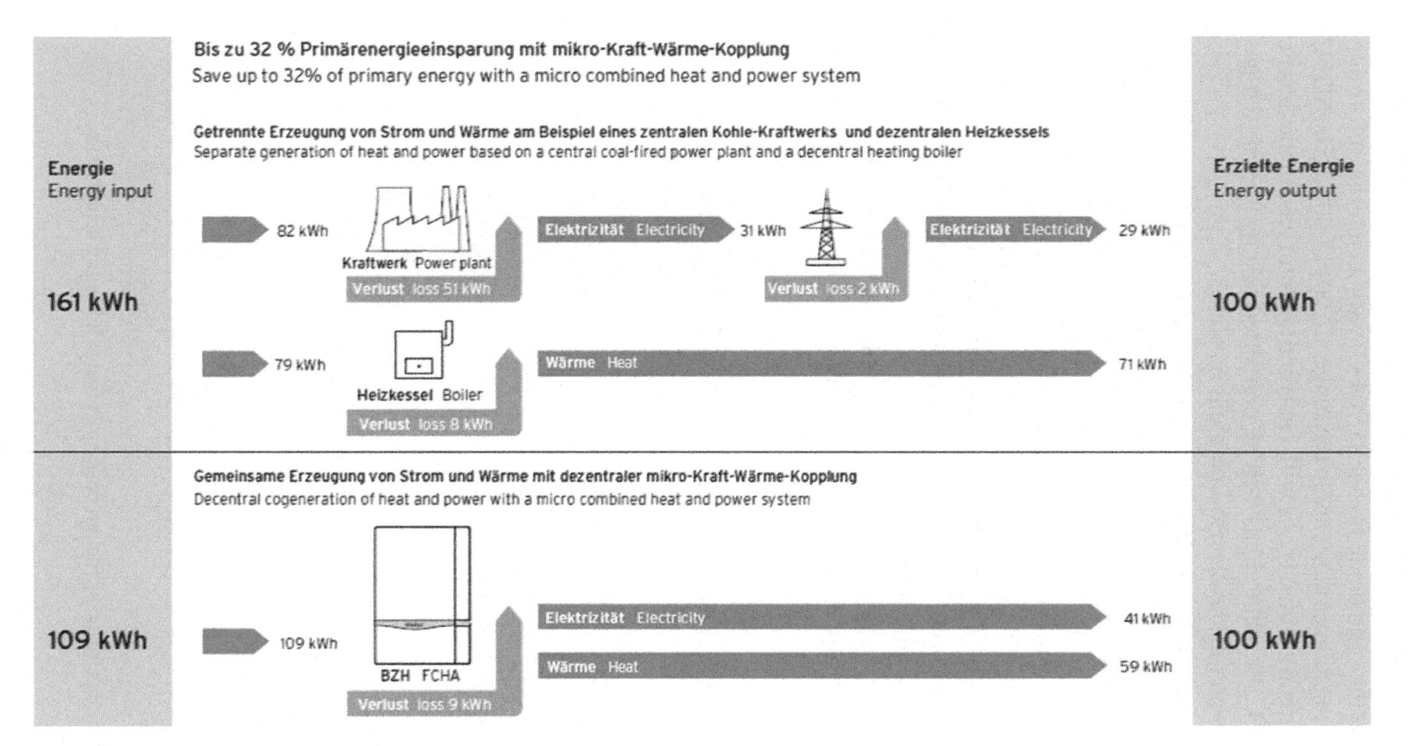

Figure 1.20 Potential for primary energy reduction with the Vaillant fuel cell heat and power unit as compared to centralized coal power and a decentralized natural-gas-fired conventional boiler.

Source: Badenhop [19].

The first one is the solid oxide fuel cell (SOFC) with an internal reforming process. Its operating temperatures are in the range of 700–1,000°C. The SOFC can be operated directly with CH_4 or with (purified) natural gas. In the anodic region of the SOFC, the natural gas will be converted to a hydrogen-rich gas with the product water of the electro-chemical reaction.

This type of fuel cell was developed in the past decade by "Ceramic Fuel Cells GmbH (CFC)"—today "SOLIDpower". The first fuel cells are on the market, particularly for nearly continuous operation due to the high temperature level.

The second one is to convert the natural gas to a hydrogen-rich fuel gas in an external reforming process and feed the fuel cell with it. This technique could also be used for low-temperature fuel cells like PEMFC.

Three different fuel cell types are possible with different paths of gas conditioning, as shown in Figure 1.21.

The process of producing hydrogen by reforming natural gas is necessary as long as a hydrogen infrastructure is not yet available in each household.

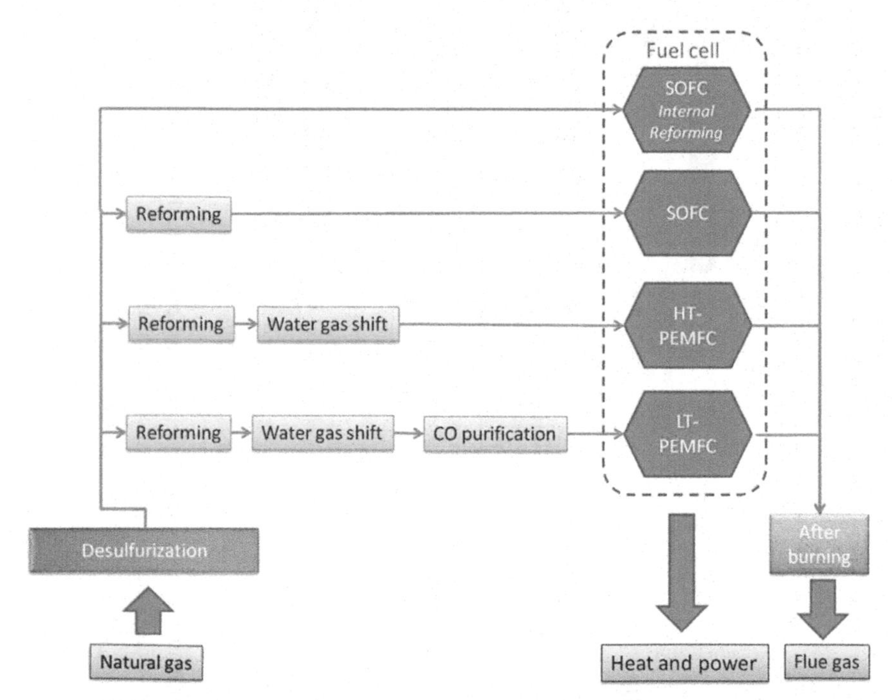

Figure 1.21 Schematic of fuel processing in fuel cell heating applications.

Source: Badenhop [19].

This concept is described in detail by Badenhop in Reference [19].

The additional reforming step naturally has some energy loss, but the waste heat from the reformer can be used as process heat internally or as additional heat for the cogeneration of heat and power. The strategic importance of the usage of natural-gas-fueled fuel cell is to introduce fuel cells into households and to increase its primary energy efficiency until hydrogen is available in each household.

In a third strategic step—when a hydrogen infrastructure is available—the reforming process is unnecessary and the primary efficiency could be increased further by at least 10%. Moreover, the system complexity could be reduced by at least 40%.

In order to prepare this technology, several German companies are involved in the "Callux" project supported by the German government and the "ene.field" project supported by the European Commission and organized in the "Initiative Fuel Cells" (IBZ). Within these projects, several hundreds of heat and power cogeneration systems are operating in public field tests. More detailed information is presented in: http://www.callux.net/home.English.html, or http://www.ibz-info.de/.

To accelerate market introduction of fuel cells for stationary application, some of these German companies formed a strategic alliance with Japanese partners: like Viessmann with Panasonic, Baxi with Toshiba, and Bosch with Aisin Seiki. In all cases, the Japanese partners have a high level of experience with the market introduction of several tens of thousands of domestic heating systems with fuel cells.

One example of a current product of these cooperations is the "Vitovalor 300-P".

The detailed configuration is shown in Figure 1.22.

The system has the following technical parameters:

A) Combined Heat and Power Unit

o Electrical output power	0.75 kW
o Thermal output	1.00 kW
o Electrical efficiency	37%
o Total efficiency	90%
o Sound power level	<49db(A)
o Maintenance interval	biennial
o Stack lifetime	60,000 h
	4,000 starts
o Power loss over lifetime	15%

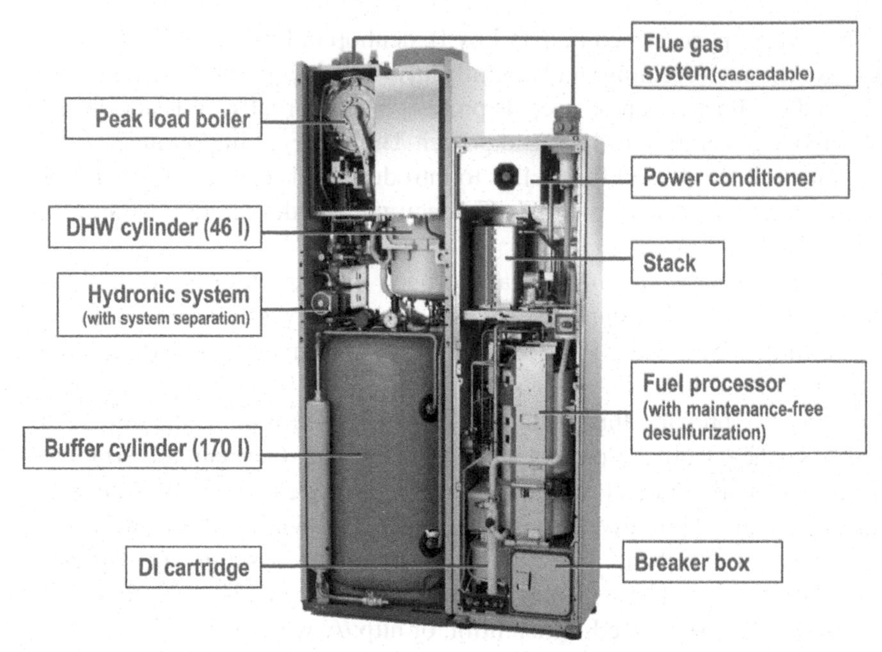

Figure 1.22 Vitovalor 300-P, System configuration.

Source: © Viessmann Werke.

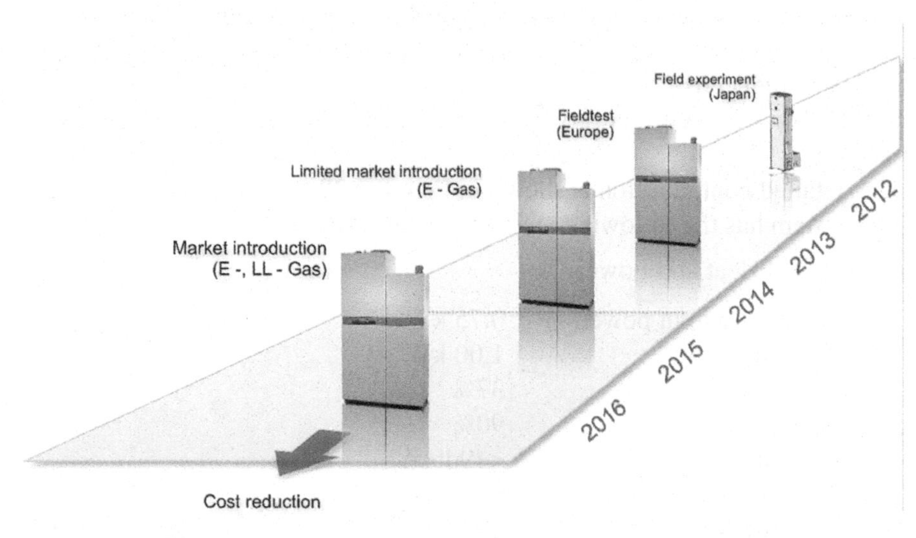

Figure 1.23 Time plan for market launch of Vitovalor 300-P.

Source: ©Viessmann Werke.

B) Peak Load Boiler
- o Thermal output power 19 kW
- o Efficiency 109%

The plan for the market penetration is shown in Figure 1.23. According to this plan, the start of market introduction in an industrial scale is to be expected in 2016.

Other manufacturers will have similar time schedules for market introduction.

Further cost reduction by increasing market scale is to be expected as well as performance improvements as experience increases.

1.9 Safety-Relevant Application

The oxygen needed for the electrochemical combustion of hydrogen in the fuel cell is used directly as pure O_2 only in some special cases (for example in submarines).

Generally, the oxygen is withdrawn from ambient air so that the exhaust air of fuel cells has a degraded O_2 concentration. For high reaction kinetics in the fuel cell, a high concentration of O_2 is necessary, and waste air is released with an O_2 concentration of ~14–15%.

As shown in Figure 1.24, the risk of ignition is nearly avoided in this concentration range.

Therefore, a preventive fire protection can be achieved using the waste air of fuel cells.

This is a very safe solution in all places where reliable fire protection of valuables, goods, and employees is required and where there are particularly high fire risks, for example, in the storage of delicate or cooled goods and in the operation of delicate computing technology.

The main applications of this technology are in the field of:

- Data centers and IT facilities;
- Warehouses and logistics centers;
- Refrigerated stores;
- Archives, museums, and libraries; and
- Chemical industry and energy storage.

In an industrial setting, fuel cells are mostly used for combined heat and power (CHP)—partly also for combined cooling, heat and power (CCHP), using an additional absorption chiller. If the latter is also used for fire protection, the fuel cell system combines four applications in a single unit.

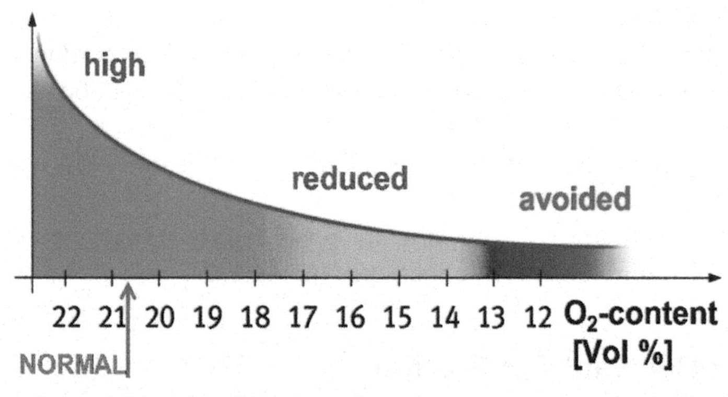

Figure 1.24 Risk of fire depending on oxygen concentration in air.

Source: N_2 telligence.

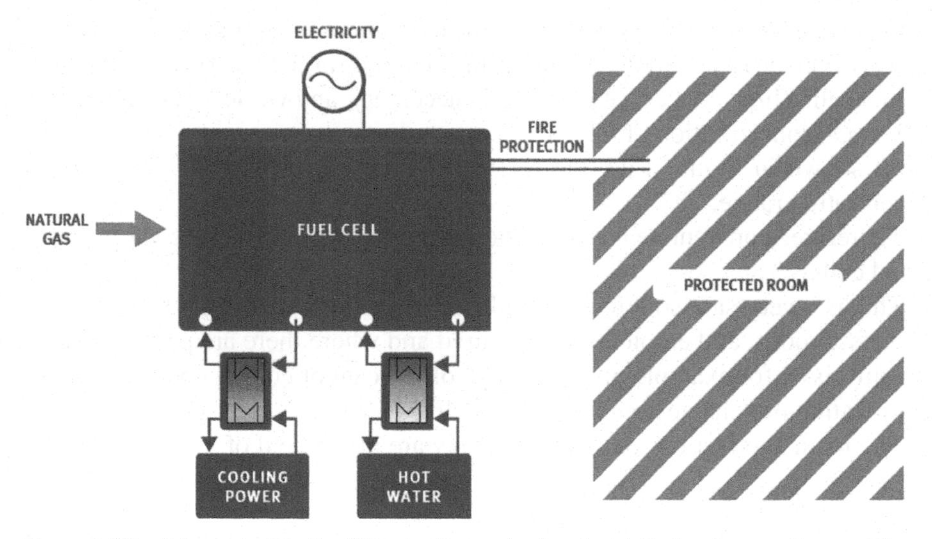

Figure 1.25 Schematical draft of "QuatroGeneration"—electricity, heating, cooling, and fire protection.

The system as a whole, named Quattro Generation, is already sold and distributed.

A schematic application concept is shown in Figure 1.25 for cogeneration of electricity, heat, cooling, and O_2-degraded air.

Table 1.3 Parameters of the Quatro generation 100 system

Electrical Power	100 kW
Voltage	400 VAC
Frequency	50 Hz
Heat Extraction	≈54 kW at 92°C or ≈40 kW at 6°C ≈54 kW at 62°C
Energy Efficiency	≈90%
O_2 Reduction in Protected Space	up to several 1,000 m^3
Energy Source	natural gas, hydrogen, biogas
Dimensions	2.2 m (W) × 6.5 m (L) × 3.4 m (H)
Weight	15.5 t

Source: http://www.N2telligence.com

The fuel gas can be natural gas, biogas (both with an additional reforming process), or hydrogen.

The technical parameters of the Quattro Generation system that is available on the market are listed in Table 1.3.

Further details of the safety-relevant application of fuel cells are reported in Reference [20].

1.10 Hydrogen Storage in Caverns

In a future sustainable energy system, one main challenge will be the volatility of primary energy carriers so that storage of energy is an indispensable factor for a basic energy supply system.

The most important issues of energy storage systems are the energy density of the stored energy carrier and the conversion efficiencies. With respect to energy densities, the storage of chemical energy (enthalpy of chemical reaction) has much higher values than the storage of potential energy. Some examples are shown in Figure 1.26.

Under the terms of storage, hydrogen has an energy content of 280 kWh/m^3. With subtraction of the conversion efficiencies of electrolysis and fuel cell, a reversible electric storage density of 170 kWh/m^3 is possible. (CH_4 has a higher storage density, but lower conversion efficiencies).

The storage of potential energies like compressed air systems (including systems with adiabatic operation) or pumped-hydrosystems have lower energy densities by a factor of 70–240. Consequently, H_2 is very suitable for the storage of large amounts of energy up to several hundreds of MWh's, as shown in Figure 1.27.

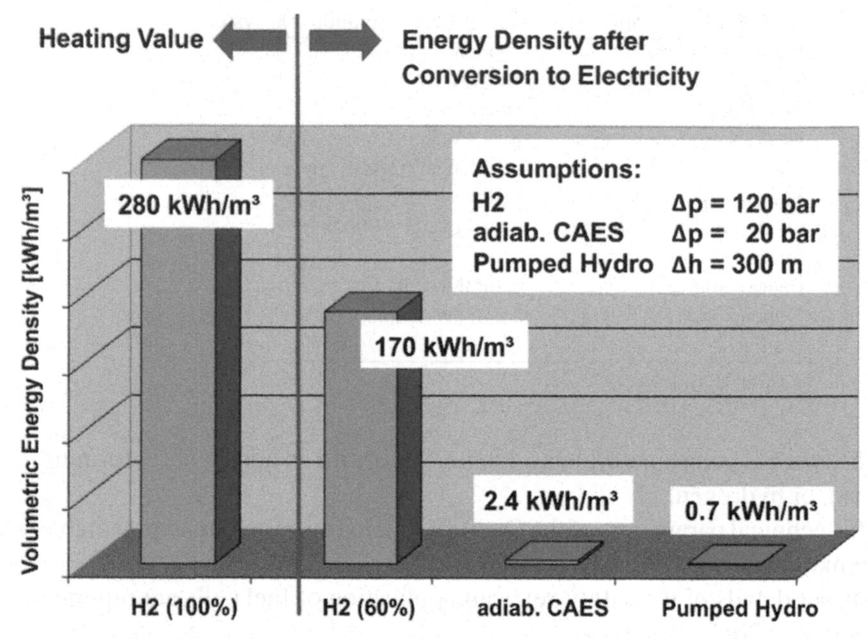

Figure 1.26 Energy densities for different energy storage systems.

Source: © KBBUT.

Figure 1.27 Comparison: Pumped hydro vs. hydrogen storage.

Source: © KBBUT.

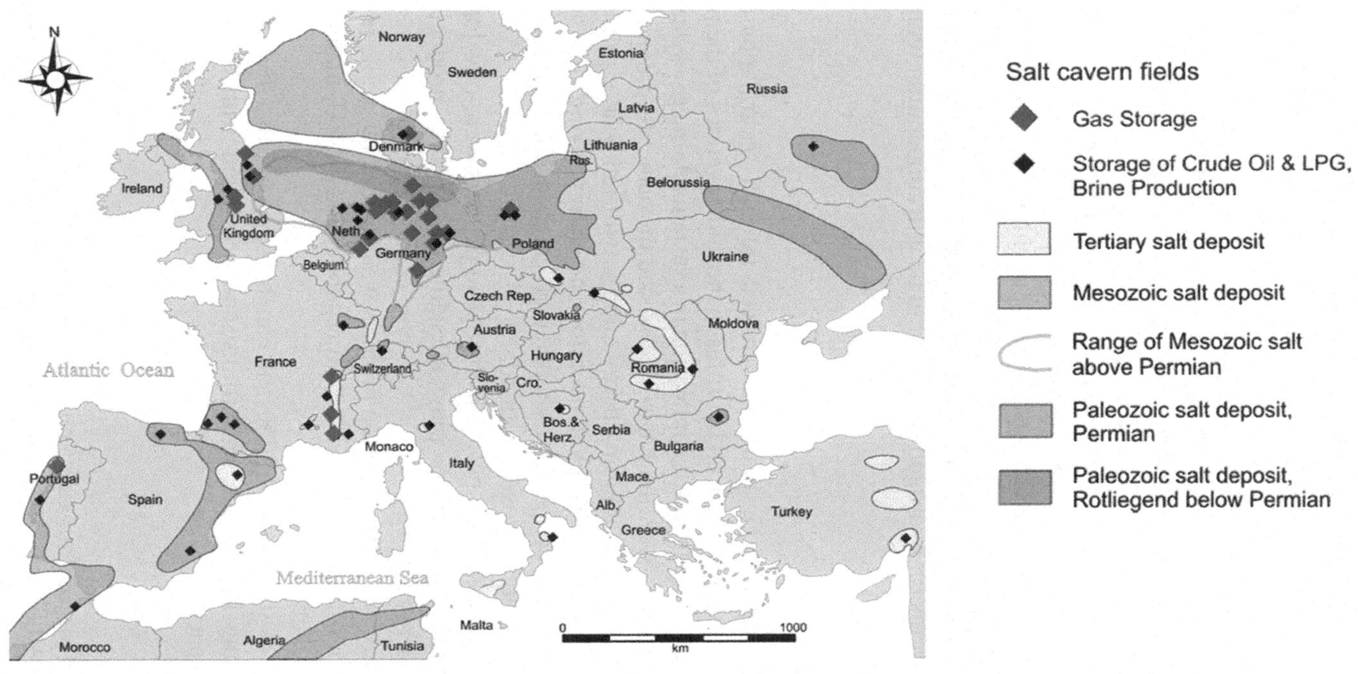

Figure 1.28 Geological map of Europe with potential for salt caverns for gas storage.

Source: © KBBUT.

In Figure 1.27, the largest pumped hydropower station in Germany (Goldisthal) is compared with a typical salt cavern field (not yet realized, but in planning).

With caverns of this type, energy storage up to TWh's will be available.

In continuous operation these salt caverns not only enable the necessary storage for stabilization of the electric grid but also the supply of hydrogen for applications like those described in the previous chapters.

A detailed investigation of the cost of different energy paths for these purposes was done by Grube and Höhlein [21].

Figure 1.28 shows the salt deposits for Europe with the potential for salt caverns for gas storage.

With this map, each country can decide to look for its own opportunities to develop salt caverns for hydrogen storage or to do it in cooperation with its neighbours.

1.11 Conclusion

Hydrogen development in Germany has moved through several phases–starting with basic research on metal hydrides in the mid-seventies of the past century to products and systems for application in a sustainable energy system.

The whole time, however, the basic idea stayed the same: to prepare an important contribution to enable an energy supply with renewable primary energies, high efficiencies of energy conversion and energy saving.

In this time, numerous cooperations between research institutes, industrial companies, public institutions were formed.

Very early on, cooperations between European partners were developed and expanded so that today Europe plays a leading role for the introduction of a sustainable energy supply.

Several research and demonstration projects were funded by public institutions, cities, regions, federal states, federal republic, and the EU, mainly in public–private partnerships.

On the basis of these cooperations, it can be expected that hydrogen will gain increasing importance in a future energy system with minimal environmental impact.

References

[1] "Neuen Kraftstoffen auf der Spur—Alternative Kraftstoffe für Kraftfahrzeuge" Published by BMFT, 1974, Verlag Gersbach und Sohn, Munic.

[2] Reilly, J. J., Wiswall, Jr., R. H. Formation and properties of iron titanium hydride. Inorganic Chemistry.1974; 13: 218.

[3] Bernauer, O., Töpler, J., Noréus, D., Hempelmann, R., Richter, D. Fundamentals and properties of some Ti/Mn-based laves phase hydrides. International Journal of Hydrogen Energy. 1989; 14(3):187–200.

[4] Töpler, J., Feucht, K. Results of a fleet test with metal hydride motor cars. Zeitschrift für physikalische Chemie. 1989; 164: 1451–1461.

[5] Bernauer, O., Buchner, H., Baier, H. German Patent P2855467.9 (1978), and US Patent 4,310,601 (1979).

[6] Töpler, J., Bernauer, O., Buchner, H. Use of hydrides in motor vehicles. Journal of Less Common Metals. 1980; 74: 385–399.

[7] Jüntgen, H., Knoblauch, K., Harder, B. Carbon molecular sieves: production from coal and application in gas separation. Fuel. 1981; 60(9): 817ff.

[8] "Alternative Energy Sources for Road Transport – Hydrogen Drive Test" Published by TÜV Rheinland e.V., ISBN 3-88585-77-8, Köln (1990).

[9] Buchner, H., Schmidt-Ihn, E., Kliem, E., Lang, U., Scheer, U. "Hydride Storage Devices for Load Leveling in Electric Power Systems" European Communities, Contract-Nr. 602-78-EHD (www.bookshop.europa.eu/.../CDNA07314ENC_001.pdf).

[10] Bernauer, O., Halene, C., Schmidt-Ihn, E., Töpler, J. "Construction and Testing of a Stationary Hydride Storage Facility" EEC-Agreement Nr. EHC-44-015-D, Final-Report 1987. (www.bookshop.europa.eu/...DE.../CDNA11445DEC_001.pdf).

[11] Peschka, W. "Flüssiger Wasserstoff als Energieträger", Springer-Verlag Wien, (1984) ISBN 3-211-81795-6.

[12] Gretz, J., Drolet, B., Kluyskens, D., Sandmann, F., Ullmann, O. Status of the hydro-hydrogen pilot (EQHHPP). Int. J. Hydr. Energy. 19(2): 169–174.

[13] Schindler, J., Wurster, R. "EURO-QUÉBEC HYDRO-HYDROGEN PILOT PROJECT [EQHHPP]- Appraisal and indications for the future" EU-Contract No. 10228-94-06 F1PC ISP D–Amendment No. 1 and 2 FINAL REPORT, April 1999.

[14] Töpler, J., Lehmann, J. (Eds.). "Hydrogen and Fuel Cell—Technologies and Market Perspectives." Springer–Verlag, Heidelberg (2015), ISBN 978-3-662-44971-4.

[15] Mohrdieck, Ch., Venturi, M., Breitrück, K., Schulze, H. "Automotive Application." in Töpler, J., Lehmann, J. (Ed.). "Hydrogen and Fuel

Cell—Technologies and Market Perspectives", Springer–Verlag, Heidelberg (2015), ISBN 978-3-662-44971-4.
[16] REGULATION (EC) No 79/2009 OF THE EUROPEAN PARLIAMENT AND OF THE COUNCIL of 14 January 2009 on type-approval of hydrogen-powered motor vehicles, and amending Directive 2007/46/EC.
[17] COMMISSION REGULATION (EU) No. 406/2010 of 26 April 2010 implementing Regulation (EC) No. 79/2009 of the European Parliament and of the Council on type-approval of hydrogen-powered motor vehicles.
[18] www.global-hydrogen-bus-platform.com/www.global-hydrogen-bus-platform.com/index.html
[19] Badenhop, T. "Fuel cells in the energy supply of households." in Töpler, J., Lehmann, J. (Eds.). "Hydrogen and Fuel Cell—Technologies and Market Perspectives", Springer–Verlag, Heidelberg (2015), ISBN 978-3-662-44971-4.
[20] Frahm, L. "Safety-relevant Application." in Töpler, J., Lehmann, J. (Eds.). "Hydrogen and Fuel Cell—Technologies and Market Perspectives", Springer–Verlag, Heidelberg (2015), ISBN 978-3-662-44971-4.
[21] Grube, T., Höhlein, B. "Costs of Making Hydrogen Available in Supply Systems Basedon Renewables." in Töpler, J., Lehmann, J. (Eds.). "Hydrogen and Fuel Cell—Technologies and Market Perspectives", Springer–Verlag, Heidelberg (2015), ISBN 978-3-662-44971-4.

2

From Zero to Present, the Hydrogen Technologies and Economy in South Korea

Byeong Soo Oh

Chonnam National University,
77 Yongbong-ro, Buk-gu, Gwangju, 61186, South Korea

2.1 History of Energy Use in Korea from 1950 to Present

In 1945, Korea was divided into Republic of Korea (ROK), which is called simply South Korea by world people, and Democratic People's Republic of Korea (DPRK), which is called North Korea. Korea was the poorest country in the world just after the Korean War from June 1950 to July 1953. Too many people were killed during the war and died from hunger and disease because of the war. Most people went to mountain to get wood for cooking and heating rooms because there were no electricity and other fuels. In the 1960s, coal briquette was used instead of wood. Some people died because the carbon monoxide from incomplete combustion of the coal permeated into heating room at night. Steam locomotives used coal for the steam turbine to get power. In the 1970s, oil was used for cooking, heating, and hot water in house. Diesel engines with oil were used for power of trains. Many cars, buses, and trucks with internal combustion engine were shown up in the streets. The first expressway road was constructed between the capital city, Seoul, and the second largest city, Busan. In the 1980s, people began to buy their own cars. Since 1990s natural gas has been used for cooking and heating in the house. Prof. Oh proposed a concept of "history of fuel and engine" as shown in Figure 2.1 at the World Hydrogen Energy Conference (WHEC) at Buenos Aires, Argentina, in 1998. Natural gas could be used in the internal engine as well as oil. From the 2000s all the city buses in ROK were changed from diesel buses to natural gas buses in order to reduce air pollution in city area. Hydrogen is the final chemical fuel and the fuel cell with hydrogen is

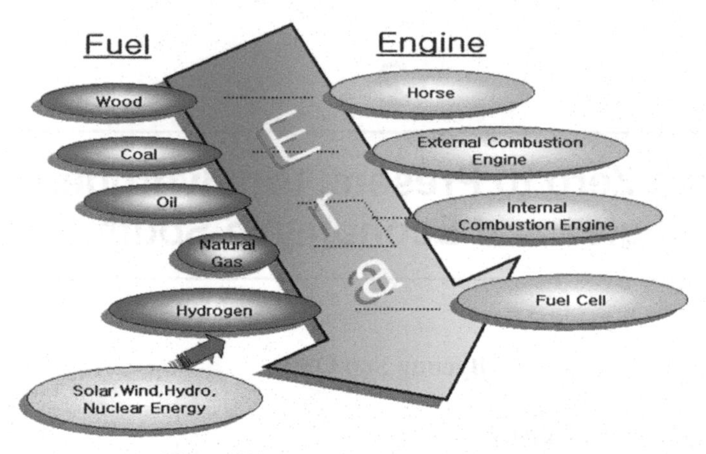

Figure 2.1 History of fuel and engine.

the new engine instead of internal combustion engine. In the 2010s, more than 100 hydrogen fuel cell cars and some fuel cell buses are operated in ROK. More than 17 hydrogen fueling stations were built in ROK. ROK is the fastest growing country in the world economically and became the world's 11th biggest economic country. Prof. Oh suggested a concept as shown in Figure 2.2 to produce hydrogen from natural energy sources and to use the hydrogen for cooking, heating, and hot water in house and for power of all the transportation machines. Usually, internal combustion engine has four, six, or eight cylinders in which fuel and air are combined to make power with CO_2 and H_2O. Human body, which is the best natural machine, has more than 75 trillion distributed cells in the body. Each cell makes energy with CO_2 and H_2O by using nutrient and oxygen in the blood as shown in Figure 2.3. Fuel cell stack in a fuel cell car has about 500 cells, which make power like engine cylinder and human body cell. In our future, the decentralized system of energy production and consumption is the most feasible way, which can be achieved by hydrogen production and consumption.

2.2 A History of Hydrogen Energy and Fuel Cell in Chonnam National University

Hydrogen and Fuel Cell Laboratory in Chonnam National University is an example of growing interest for hydrogen energy and fuel cell vehicle in Korea. The advisor, Prof. Oh, in the laboratory was born in 1953 at the end of Korean War and has all the experience of the Korean energy history such

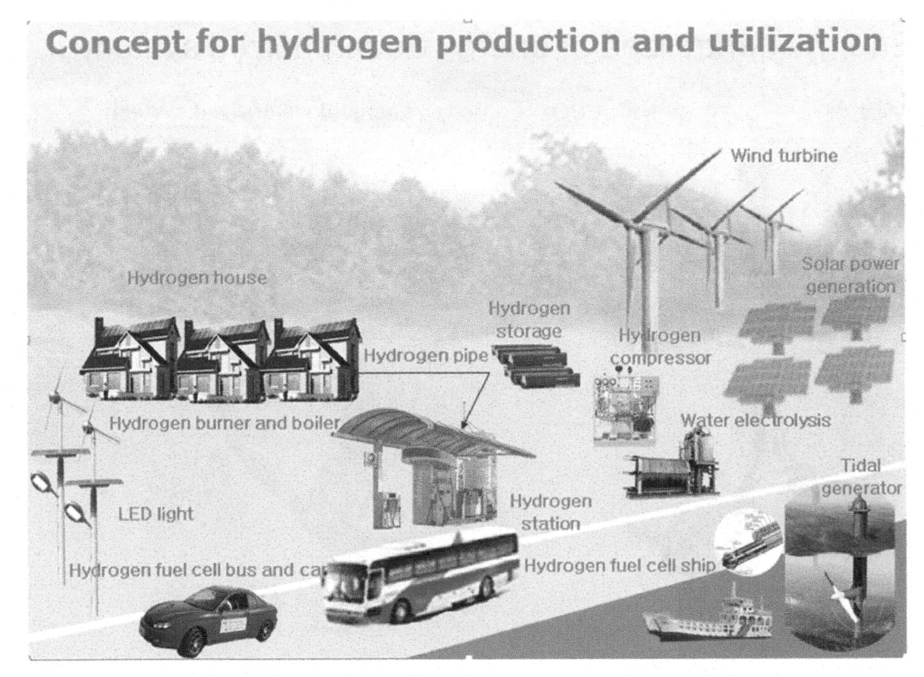

Figure 2.2 Concept for hydrogen production and utilization.

as wood, coal, oil, natural gas, and hydrogen energy. He met Dr. Kil Hwan Kim, who was the Founding President of the Korean Hydrogen Energy Society (KHES), which is now called "The Korean Hydrogen and New Energy Society (KHNES)". He gained a membership from the International Association for Hydrogen Energy (IAHE) in 1982. "A study on hydrogen generation by using water electrolysis and solar cell" was published in 1982 as his first paper about hydrogen energy. Dr. Kim and he with two other persons had tried to found the KHES in 1983. Actually, the KHES was founded in 1989. After he received his PhD degree from the State University of New York at Stony Brook, USA, in 1989, he attended the World Hydrogen Energy Conference (WHEC 1990) in Hawaii, USA, for the first time in his life because Dr. Kim introduced him about the conference and recommended to attend together. Since then until now he has attended all the WHECs because those provided him with a plenty of information about hydrogen energy and fuel cell. At first, he made a "Hydrogen Energy & Power Laboratory" to make a non-pollution hydrogen car by using micro gas turbine with combustion of pure hydrogen and pure oxygen in combustion chamber. In the 1980s, most researchers in the world

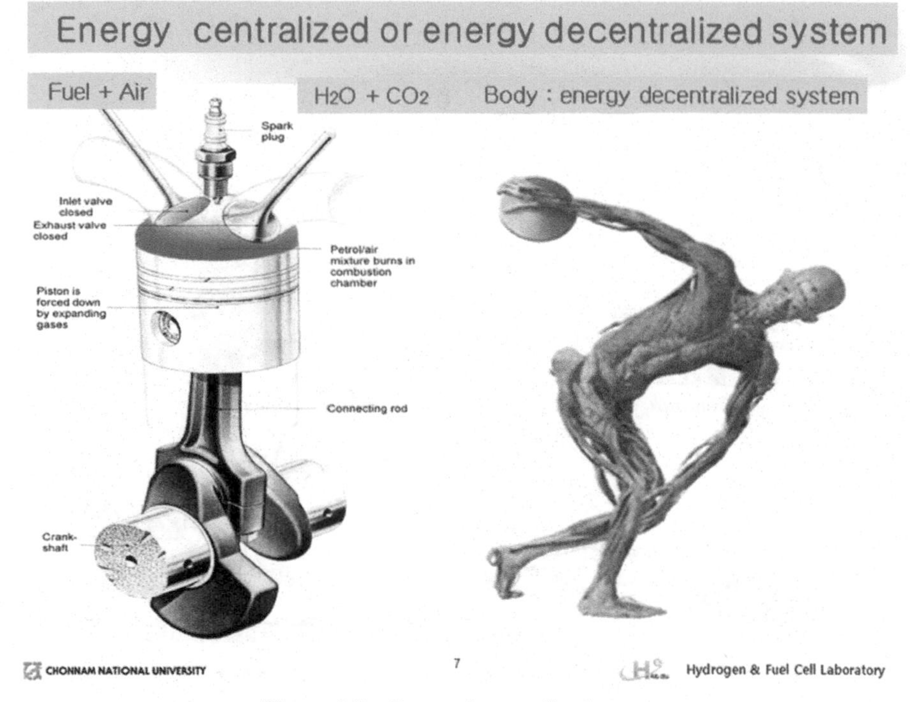

Figure 2.3 Energy decentralized system.

made hydrogen cars with internal combustion engine and liquid hydrogen tank. Many hydrogen cars from USA and Japan were exhibited at WHEC 1990 in Hawaii. He went to Prof. Veziroglu at Miami University, USA, in 1997 as a visiting professor for one year. He read there numerous papers about fuel cell vehicles and discussed about the articles with Prof. Veziroglu on every Saturday. After he returned to Korea in 1998, he changed his laboratory's name to "Hydrogen and Fuel Cell Laboratory" to study about fuel cell instead of heat engine such as hydrogen gas turbine. In 2000, he made an "Experimental equipment for hydrogen energy production and utilization," as shown in Figure 2.4, and got a patent. It has been used for the basic experiment of energy conversion to junior students of the Department of Mechanical Engineering in Chonnam National University every year until now. Students can understand the conversion efficiencies of electrolyser and fuel cell, and about solar energy, hydrogen energy, and electric energy. In 2001, he established a "Club for Hydrogen Fuel Cell Vehicle (CHFCV)" of undergraduate students. Usually, making a new machine is quite difficult and operating the new machine successfully is much more difficult. He never saw any real fuel cell stack

Figure 2.4 Experimental equipment for hydrogen energy production and utilization.

before he and his student made a Polymer Electrolyte Membrane Fuel Cell (PEMFC) stack with 15 cells in 2002, as shown in Figure 2.5, and operated it successfully. Membrane Electrode Assembly (MEA) was made by using a Hot Press which attached Pt/C powder and carbon cloth on polymer membrane according to the description of other's papers. Later, the student got a doctor's degree and was known as the first doctor of PEMFC in the field of Mechanical Engineering in Korea. The doctor has been working for the development of fuel cell car in Hyundai Motor Company. The CHFCV made fuel cell cars with their own PEMFCs almost every year by using some fund from a project of the Department of Mechanical Engineering. One of the fuel cell cars was made in 2003 and is shown in Figure 2.6. In 2004, he with a master's course student developed a hydrogen catalytic burner, as shown in Figure 2.7, which used platinum catalyst like in PEMFC, and got a patent to be used in house. In 2005, he received a fund from the Director, T. Nejat Veziroglu, of UNIDO-ICHET (United Nations Industrial Development Organization—International Center for Hydrogen Energy Technologies) to make a proposal of "Hydrogen Fuelled Vehicles Project, South Korea". Two books of engineering report and funding proposal were reported to the UNIDO-ICHET. If the project were supported from Korean central government and local governments, the history of Korean hydrogen energy and fuel cell vehicles could be changed

Figure 2.5 The first PEMFC stack in Chonnam National University.

Figure 2.6 Hydrogen fuel cell car in Chonnam National University.

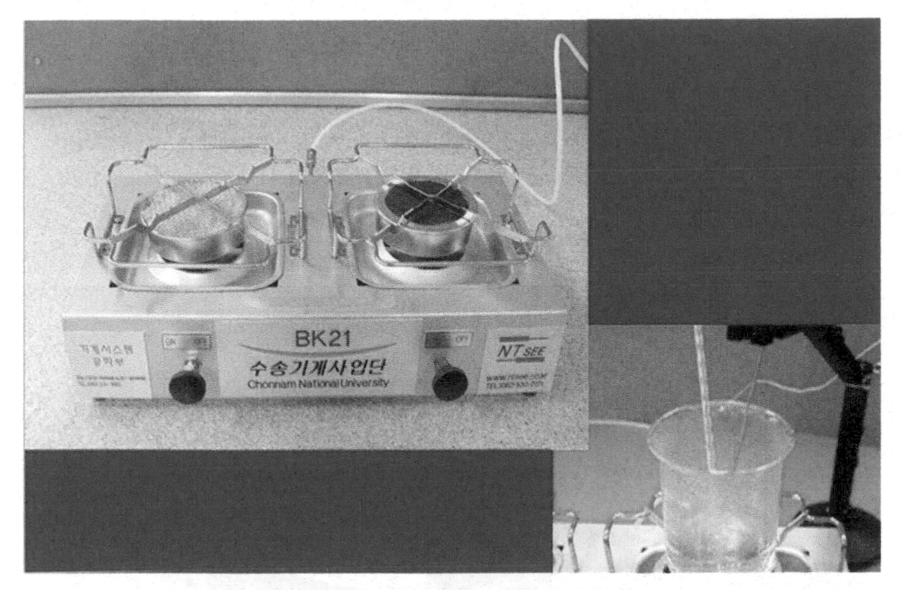

Figure 2.7 Hydrogen catalytic burner.

more rapidly. In the Department of Mechanical Engineering, the subjects of "Hydrogen Energy Application" in 2004 and "Fuel Cell Power System" in 2006 for graduate course and "Fuel Cell Vehicle" in 2005 for undergraduate course were added in the curriculum and have been taught by him until now. Since 2006 he has been a member of the Board of Directors in the International Association for Hydrogen Energy (IAHE). In 2008, he bought an old and nice looking red car with internal combustion engine. He and many students got rid of the internal combustion engine and installed PEMFC stack, hydrogen tank, BLDC motor and batteries to make a fuel cell car as shown in Figure 2.8. The fuel cell car could be operated by voice recognition and touch screen monitor for stop, start, and 5 step speeds. The next year, they got the first prize, Award of the Minister of Knowledge Economy, during an international science and technology competition of renewable energy utilization. He as a mechanical engineer was concerned about the mechanism of hydrogen fuel cell car. Many graduate students studied how to control and make fuel cell car as well as fuel cell itself with Balance of Plant (BOP), and how to save regenerative energy as shown in Figures 2.9 and 2.10. In 2010, he founded a "Hydrogen and Fuel Cell Research Institute" in Chonnam National University. From 2008 to 2014 he became the chairman of the World Hydrogen Energy Conference (20 WHEC 2014) held at Gwangju City, his hometown, Korea.

Most students who became bachelors, masters, and doctors from his laboratory have studied toward hydrogen energy and fuel cell vehicle so that they believe riding an autonomous hydrogen fuel cell car in the near future. Finally, five professors from Chonnam National University with their graduate students and some family members attended at the 21st WHEC 2016, Zaragoza, Spain in June, 2016.

Figure 2.8 Hydrogen fuel cell car 1 in Chonnam National University.

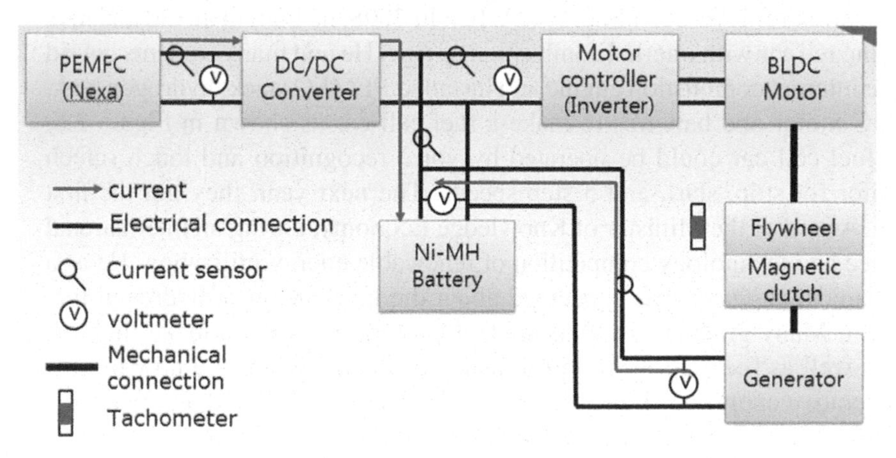

Figure 2.9 Schematic diagram of regenerative braking system.

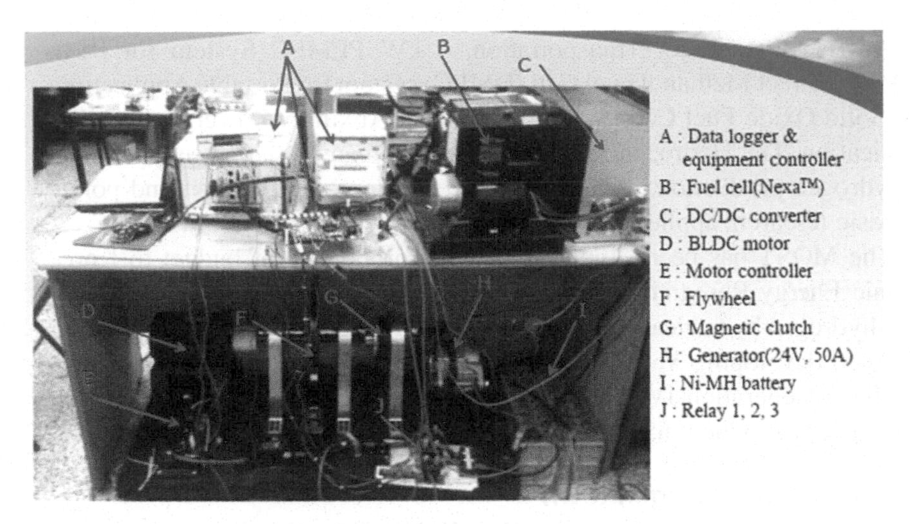

Figure 2.10 Regenerative braking system.

2.3 Activities for Hydrogen and Fuel Cell in ROK

There are many companies, research institutes, universities, and associations, which are related to hydrogen energy and fuel cell in Korea.

2.3.1 Hydrogen and Fuel Cell Activities Supported by Government

Korean government supported US$5 million for hydrogen and US$35 million for fuel cell from 1988 to 2002. Two departments of central government, which were called Ministry of Commerce, Industry and Energy (MOCIE) and Ministry of Science and Technology (MOST), supported two big projects for hydrogen and fuel cell. The MOCIE supported National RD&D Organization for Hydrogen and Fuel Cell for 5 years from 2004 to 2008 and the MOST supported 21st Frontier Hydrogen Energy R&D Program for about 10 years from 2003 to 2013. The organization had subcommittees such as Hydrogen Infrastructure, Hydrogen Technology, Fuel Cell for Power Plant, Fuel Cell for Transportation, Portable Fuel Cell, Fuel Cell for Residential Power Generation (RPG) and Dissemination. Their activities focussed on Hydrogen Energy R&D Program, Hydrogen Production from Nuclear Energy, Development and Demonstration of Hydrogen Station, High Pressure Vessel for Hydrogen Storage, Hydrogen Codes, Standards and Safety, Development of Molten Carbonate Fuel Cell (MCFC) System for Stationary Application, Development of PEMFC System for Transportation, 3 kW PEMFC System for RPG, PEMFC, Direct Methanol Fuel Cell (DMFC) system for Portable Application, and Solid Oxide Fuel Cell (SOFC) system for Auxiliary Power Unit (APU) Application. The Hydrogen R&D Program had many research subjects such as hydrogen production, hydrogen storage, hydrogen utilization and policy for basic research, applied research and demonstration.

The MOST has been supporting a large proportion of budget to Korea Atomic Energy Research Institute (KAERI) (http://www.kaeri.re.kr) to produce hydrogen by nuclear energy for 16 years from 2004 to 2020. The KAERI has been developing Very High Temperature gas-cooled Reactor (VHTR), which is a Generation-IV reactor for the mass production of hydrogen as well as electricity production.

2.3.1.1 Ulsan hydrogen town

Ministry of Trade, Industry and Energy (MOTIE) (http://english.motie.go.kr), whose former name was MOCIE, and Ulsan City Government have been supporting Ulsan Hydrogen Town from 2012 to 2018 (http://h2town.utp.or.kr). About 67% of national hydrogen production is produced as by-product gas from the chemical industries in Ulsan industrial complexes. The price of by-product hydrogen is the cheapest among other hydrogens, which are produced by using natural gas reformer, wind turbine, or solar cell. There are 140 apartments with 1 kW fuel cell each, 9 houses and offices with 5 kW fuel cell each, and 1 dormitory with 10 kW fuel cell. The fuel cells were made by many companies such as GS Caltex, Fuel Cell Power, Hyosung, and Hyundai Hysco. Ulsan Techno Park manages the town and SPG company checks the safety of all system.

2.3.1.2 Hydrogen stations

Most of the 17 hydrogen stations, as shown in the Figure 2.11, were built by support of the government from 2001 to 2014. The MOCIE or MOTIE supported the most budget of the installing stations as development stage until 2013. From 2014 hydrogen stations as dissemination stage have been installed by the half support of total installing cost from the Ministry of Environment (ME) (http://www.me.go.kr). Still any company which wants to install the hydrogen station needs more support from the local government.

2.3.2 The Hydrogen and New Energy Society

Dr. Kil Hwan Kim founded "The Korean Hydrogen Energy Society (KHES)" (http://www.hydrogen.or.kr) in July, 1989, for the purpose of cooperating among industrial, academic, governmental, and ordinary people to use the

	location	company	year	Type	Fund	Remark
1	Hwasung	HMC	2001	Truck-In	HMC	
2	Yongin	HMC	2005	Truck-In	HMC	-
3	Daejeon	Kier	2006	NG reforming	MEST	Prontier
4	Seoul	GS Caltex	2007	Nathae reforming	MKE	Demo.
5	Incheon	KOGAS	2007	NG reforming	MKE	Demo.
6	Daejeon	SK	2007	LPG reforming	MKE	Demo.
7	Seoul	KIST	2008	Truck-In	MKE	monitoring
8	Hwasung	HMC	2008	Truck-In	HMC	-
9	Ulsan	Dongdeok Gas	2009	Truck-In	MKE	monitoring
10	Yeosu	SPG Chemical	2009	Truck-In	MKE	monitoring
11	Jeju	HMC	2010	Eletrolysis	MKE	monitoring
12	Seoul	Seoul City	2010	LFG reforming	Seoul	-
13	Seoul	HMC	2010	Truck-In	HMC	-
14	Buan	Kier	2012	Eletrolysis	MKE	-
15	Ulsan	Dongdeok Gas	2013	Truck-In	MKE	Social Demo.
16	Daegu	EM Korea	2013	Eletrolysis	Daegu	-
17	kwangju	-	2014	-	ME	-

Figure 2.11 Hydrogen stations in Korea.

ideal energy of hydrogen instead of using fossil fuel. The first Korea–Japan Joint Symposium on Hydrogen Energy was held at Seoul National University, Seoul, Korea in 1991. The symposium had been held in every odd year between Korea and Japan. It became Asian Hydrogen Energy Conference by including China and other countries, which had been continued until 2009. Sometime it will be continued because of increasing demand of hydrogen energy and fuel cell field. In 2002, the KHES was changed to "The Korean Hydrogen and New Energy Society (KHNES)" to get broader field.

2.3.3 Korea Hydrogen Industry Association

Korea Hydrogen Industry Association (KHIA) (http://h2.or.kr) was founded in 2014 for the purpose of cooperation among mainly industrial companies with some universities, research centers, and associations. The aim is to communicate related information about fuel cell vehicles, domestic fuel cell, fuel cell power plant, and hydrogen energy economy.

2.3.4 Korea Institute of Science and Technology

The government funded Korea Institute of Science and Technology (KIST) (http://www.kist.re.kr), found in 1966, is the first science and technology research institute in Korea and has been playing a leading role of national research. From the early 1990s, KIST has been studying actively about hydrogen energy and fuel cell such as hydrogen gas detector, metal hydride, cooling system for liquefied hydrogen, liquid hydrogen tank, hydrogen-oxygen flame, MCFC, PEMFC, and SOFC. KIST studied also about DMFC, Microbial Fuel Cell (MFC), fuel cell vehicle, and hydrogen production by using solar cell or by reforming of methane, etc.

2.3.5 Korea Institute of Energy Research

The Korea Institute of Energy Research (KIER) (http://www.kier.re.kr) was established as the name of "The Korea Institute Energy Conversion (KIEC)" in 1977. The KIER's major aim is to develop energy efficient technology and to make nation's energy policy. Recently, the KIER has been studying about how to reduce the use of fossil fuels such as oil, coal as well as natural gas and to increase renewable energy sources such as solar energy by photovoltaics and thermal collectors, wind energy, and bioenergy with hydrogen energy, fuel cell, etc. There are many research reports about hydrogen energy and fuel cell since the early 1980s. Hydrogen and Fuel Cell Center for Industry, Academy and Laboratories was established in 2011 for the special purpose of hydrogen and fuel cell research.

2.3.6 Hyundai Motor Company

- Hyundai Motor Company was founded in 1967 (http://www.hyundai.com). The company is one of the best companies of producing hydrogen fuel cell vehicles in the world. There is too much information available of Hyundai's history about hydrogen and fuel cell vehicles. A small part of the information is introduced in this chapter.
- Korea's first automotive fuel cell was developed in 1999. Korea's first fuel cell electric vehicle Santa Fe was developed in 2000. The world's first ultra high pressure hydrogen storage system for fuel cell electric vehicles was developed in 2003. Tucson fuel cell electric vehicle was developed in 2004. The 3rd generation fuel cell concept car i-Blue was unveiled in 2007. World's 1st mass production of ix35 (Tucson) fuel cell electric vehicles was started in 2013 as shown in Figure 2.12.
- Tucson ix35 fuel cell car's specifications are as follows. It has metal bipolar plate, Li-polymer battery, maximum power of 100 kW (+24 kW battery), maximum torque of 300 Nm, fuel tank of 5.64 kg H_2 (700 bar), driving range of 594 km, acceleration (0 $\rightarrow$ 100 km/h) of 12.5 sec, and maximum speed of 160 km/h.
- Hyundai made a plan as shown in Figure 2.13 and tried to follow the plan. But the environment in the field of hydrogen energy, hydrogen refueling station, and fuel cell cannot support the plan quickly nowadays.

- The second domestic social validation of 100 Hyundai's fuel cell vehicles was tested in Seoul and Ulsan cities between 2010 and 2013, as shown in Figure 2.14.

Figure 2.12 The world's first mass production of fuel cell car.

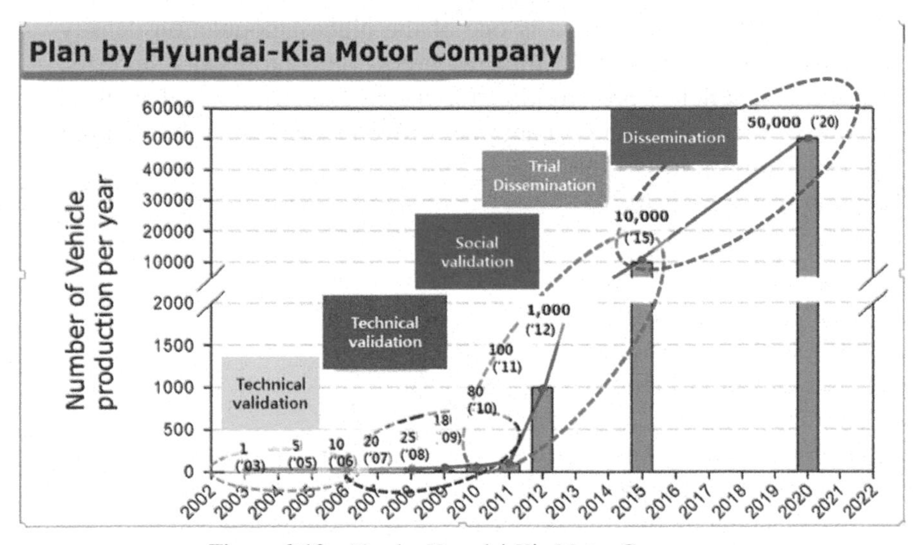

Figure 2.13 Plan by Hyundai-Kia Motor Company.

Figure 2.14 Social validation of fuel cell vehicle.

2.3.7 List of Many Universities, Companies, and Institutes Which Related Hydrogen and Fuel Cell

Human resource in Korea is significantly important to develop and improve the scientific field because Korea is one of the highest population density countries and is a poor country of natural resources. Many professors, students, researchers, and officials are working for the hydrogen and fuel cells in various universities, companies, many local governments as well as central government and institutes as follows.

Major universities are as follows:

- Ajou University
- Changwon National University
- Cheju National University
- Chonbuk National University
- Chonnam National University
- Chosun University
- Chungbuk National University
- Chungnam National University
- Daegu Gyeongbuk Institute of Science & Technology (DGIST)
- Dankook University
- Dong-A University

- Dongshin University
- Ewha Womans University
- Gwangju Institute of Science and Technology (GIST)
- Gyeongsang National University
- Hanbat National University
- Hanyang University
- Hongik University
- Hoseo University
- Information and Communications University
- Incheon National University
- Inha University
- Jeonju University
- Korea Advanced Institute of Science and Technology (KAIST)
- Kangwon National University
- Keimyung University
- Kongju National University
- Konkuk University
- Kookmin University
- Korea Aerospace University
- Korea Maritime University
- Korea National University of Transportation
- Korea University
- Kumoh National Institute of Technology
- Kyungil University
- Kyunghee University
- Kyungpook National University
- Mokpo National University
- Myongji University
- Pohang University of Science and Technology (POSTECH)
- Pukyong National University
- Pusan National University
- Sejong University
- Seoul National University
- Seoul National University of Science and Technology
- Sogang University
- Soongsil University
- Sunchon National University
- Sungkyunkwan University
- Ulsan National Institute of Science and Technology

- University of Ulsan
- Woosuk University
- Yeungnam University
- Yonsei University

Related companies are as follows:

- Posco Energy http://eng.poscoenergy.com
- Doosan Fuel Cell http://www.doosanfuelcell.com
- S-Fuel cell http://www.s-fuelcell.com
- Samsung SDI http://www.samsungsdi.com
- Hyundai Hysco http://www.hysco.com
- HYOSUNG http://www.hyosung.com
- KD NAVIEN http://en.kdnavien.com
- LIG Nex1 http://www.lignex1.com
- PRO-POWER http://www.propower.co.kr
- KOLON INDUSTRIES http://www.kolonindustries.com
- Hankook Tyre http://www.hankooktire.com
- F Cell Tech http://www.fcelltech.com
- G Philos http://www.g-philos.co.kr
- JNTG http://thejnt.com
- DAEJOO ELECTRONIC MATERIALS http://www.daejoo.co.kr
- HEESUNG CATALYSTS CORP. http://www.hscatalysts.com
- HANCHANG http://www.hanchem.com
- Dajung Chemicals and Metal http://www.daejungchem.co.kr
- Chang Sung Corporation http://www.changsung.com
- Autoen http://www.autoen.co.kr
- CNL Energy http://cnl.co.kr
- SEBANG GLOBAL BATTERY http://www.gbattery.com
- SK http://www.sk.co.kr
- KOGAS http://www.kogas.or.kr
- Kwang Shin Machinery http://www.kwangshin.com
- Deokyang http://www.deokyang.com
- Elchemtech http://www.elchemtech.com
- EM KOREA http://www.yesemk.com

Related institutes are:

- KITECH http://www.kitech.re.kr
- Creative Economy Innovation Center in Gwangju http://ccei.creative korea.or.kr/gwangju

- KATECH http://www.katech.re.kr
- KEPRI http://www.kepri.re.kr
- KAERI http://www.kaeri.re.kr
- KIMM http://www.kimm.re.kr
- Daegu Techno Park http://www.ttp.org
- Ulsan Techno Park http://www.utp.or.kr
- KGS http://www.kgs.or.kr
- Chungnam Techno Park http://www.ctp.or.kr

2.4 World Hydrogen Energy Conference 2014

International Association for Hydrogen Energy (IAHE) was founded in 1974 by the Founding President, T. Nejat Veziroglu. The first World Hydrogen Energy Conference (WHEC) was held at Miami, USA, in 1976 by IAHE. Since then, WHECs have been held in even years at different locations around the world as shown in Figure 2.15. IAHE organized World Hydrogen Technologies Conventions (WHTC) in odd years and the first WHTC was held in Singapore in 2005.

WHEC 2014 was proposed at the Board Meeting of the IAHE when WHEC 2008 was held in Brisbane, Australia. Hydrogen energy and fuel cell were main subjects to be selected as the title of 20th WHEC 2014 at Gwangju City, Republic of Korea, as shown in Figure 2.16. Hydrogen and fuel cell seem like needle and thread. Hydrogen energy leads new paradigm of the world energy system and fuel cell follows to produce power from the hydrogen energy.

"Creating a Healthy and Peaceful Hydrogen Energy World" was selected as the conference theme of the WHEC 2014, as shown in Figure 2.17. If every country makes hydrogen by using water electrolysis with electricity produced from all natural energy sources such as wind, solar, hydraulic, and geothermal energies, the hydrogen energy with fuel cell is used as in energy independent country. Then the world will become peaceful without any energy-related war and people will live long with healthy condition in the environment of no air pollution.

Some board members of IAHE at WHEC 2008 expected Prof. Veziroglu's 90th anniversary of birth at WHEC 2014. He attended at WHEC 2014 with good health condition and celebrated his 90th anniversary as shown in Figures 2.18 and 2.19. Furthermore, they suggested celebrating his 100th anniversary at the 25th WHEC 2024, which would be hopefully held at Miami, USA, where the first WHEC was held and he has been living for his life. The average life expectancy of people in South Korea is 82 years now. Most

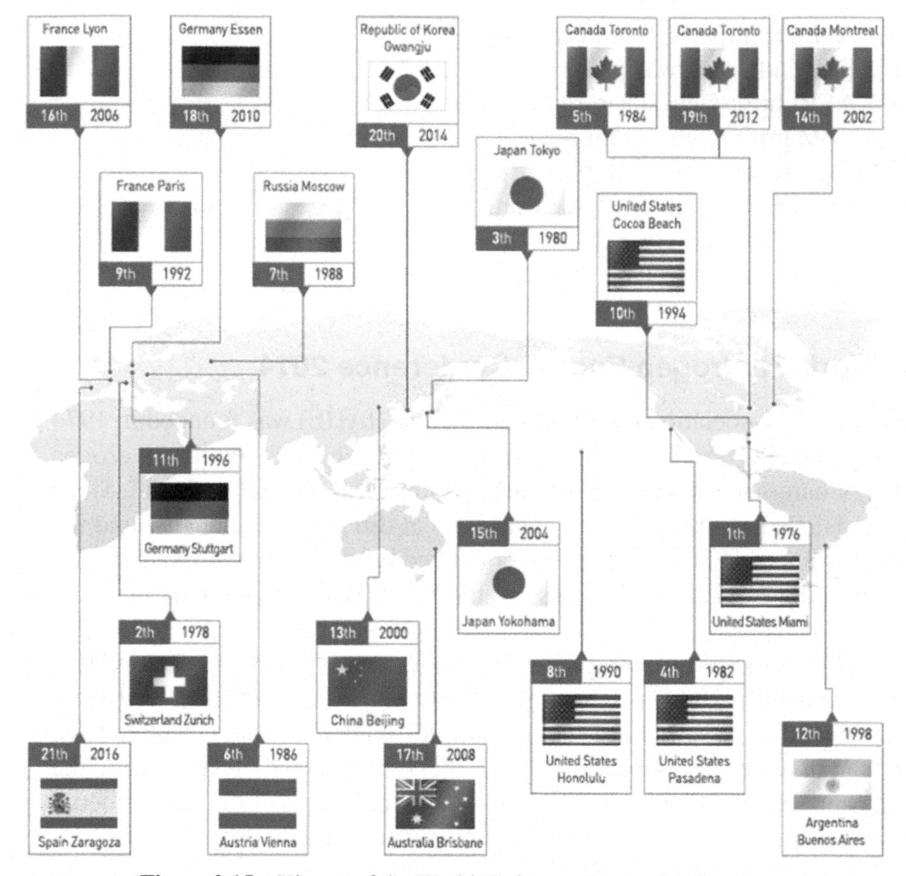

Figure 2.15 History of the World Hydrogen Energy Conference.

people would like to live longer with good health condition without any illness. Making hydrogen world will quickly improve our environment by making it clean and the result will be people's good health. On behalf of the hydrogen world, we wish Prof. Veziroglu a long, healthy, happy, and productive life as well as world people's lives.

The first commercial sale of the Hyundai fuel cell vehicle was made to the Gwangju City during the opening ceremony of the WHEC 2014. The city built one hydrogen fueling station and is operating five fuel cell cars and one fuel cell bus. As a result of the conference, the Korean Government and Hyundai Motor Company declared Gwangju City as a Hub City for Hydrogen Fuel Cell Vehicles. All the participants at WHEC 2014 received a USB, which included 450 full papers, posters, and some presentation materials. Many exhibitors participated at WHEC 2014, as shown in Table 2.1.

20WHEC2014

the 20th World Hydrogen Energy Conference
20WHEC2014 in Gwangju, Korea
June 15 - 20, 2014
http://whec2014.com

Hydrogen Energy
and Fuel Cell

E-mail : info@whec2014. com

Figure 2.16 Title of the World Hydrogen Energy Conference 2014.

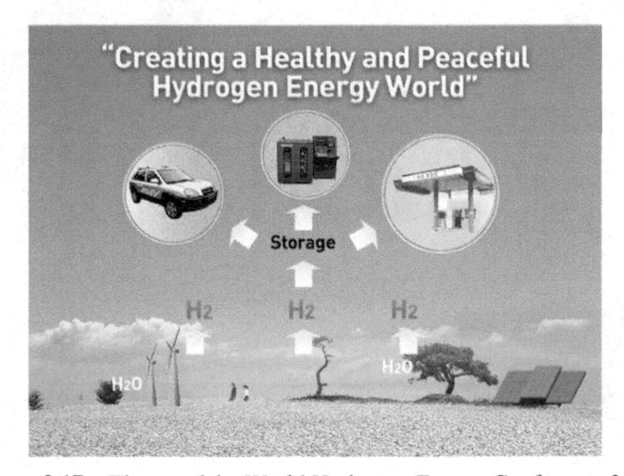

Figure 2.17 Theme of the World Hydrogen Energy Conference 2014.

Figure 2.18 Honoring Dr. Veziroglu's 90th Anniversary at the 20th WHEC 2014 Gala Dinner. (*At the Back:* Micheal-Ann & Emre A. Veziroglu, T. Nejat & Ayfer Veziroglu and Byeong Soo Oh. *At the Front:* Delmy, Derya, & Lili F. Veziroglu)

Figure 2.19 Prof. Veziroglu with some performers and participants at WHEC 2014.

Table 2.1 List of exhibitors at WHEC 2014

No	Name of Exhibitor	Nation
1	Chonnam National University	Republic of Korea
2	Elchemtech Co., Ltd.	
3	EM Korea Co., Ltd.	
4	Exco(Green Energy Expo 2015)	
5	Gwangju Metropolitan City	
6	Gwangju Federation of Handcrafts Cooperatives	
7	Gwangju Institute of Green-Car Component Industry Advancement	
8	HN Power Co., Ltd.	
9	Horizon Fuel Cell Korea	
10	Hustec Inc.	
11	Hyundai Motor Group	
12	Il Jin Composites	
13	Innowill Co., Ltd.	
14	JNTG	
15	Jung Won Commerce	
16	Sainergy Tech (USA), Inc.	
17	Korea Institute of Science and Technology	
18	Korea Tasuno Co., Ltd	
19	MS ENG	
20	Pro Power Co., Ltd.	
21	Ruby Battery Co., Ltd	
22	Sejong Industrial Co., Ltd.	
23	AECL's	Canada
24	Ballard	
25	BC Government	
26	Canadian Hydrogen & Fuel Cell Association	
27	Government of Canada	
28	Greenlight Innovation	
29	Hydrogenics Corporation	
30	EnergieAgentur.NRW	Germany
31	Forschungszentrum Jülich GmbH	
32	Linde	
33	NOW Nationale Organization Wasserstoff- und Brennstoffzellen technologie GmbH	
34	Peter Sauber Agentur Messen und Kongresse GmbH	
35	Springer-Verlag GmbH	
36	WEH GmbH Gas Technology	
37	H-TEC EDUCATION GmbH	
38	WORLD OF ENERGY SOLUTIONS	
39	International Association for Hydrogen Energy	USA
40	PDC Machines	

(Continued)

Table 2.1 Continued

No	Name of Exhibitor	Nation
41	Chiyoda Corporation	Japan
42	Tatsuno Corporation	
43	Whole Win (Beijing) Materials Sci. & Tech. Co., Ltd.	China
44	Wuhan WUT New Energy Co., Ltd	
45	AFC Energy	United Kingdom
46	Aragon Hydrogen Foundation	Spain
47	Norwegian Corner	Norway
48	AREVA H2-Gen	France

References

[1] Oh BS. Foreword: Honoring the 90th Anniversary of T. Nejat Veziroglu. *International Journal of Hydrogen Energy*. 2015; 40 (35): 11343–11345.

[2] http://www.iahe.org: International Association for Hydrogen Energy.

[3] Hydrogen Energy and Fuel Cell, 20WHEC 2014, *Proposal for Hosting the 20th World Hydrogen Energy Conference*, June 2008.

[4] http://www.index.go.kr

PART II

Hydrogen Energy in Emerging Markets

3

Hydrogen Technologies in The Czech Republic

Karel Bouzek[1], Martin Paidar[1] and Karin Stehlík[2,3]

[1]Department of Inorganic Technology, University of Chemistry and Technology Prague, Technická 5, 166 28 Prague 6, The Czech Republic
[2]Centrum výzkumu Řež, s.r.o., Hlavní 130, 250 68 Husinec-Řež, The Czech Republic
[3]Czech Hydrogen Technology Platform, Hlavní 130, 250 68 Husinec-Řež, The Czech Republic

3.1 Hydrogen Technologies in The Czech Republic—History, Presence, and Perspectives

3.1.1 History

Although not really expected, activities in the field of hydrogen technologies date in The Czech Republic more or less 5 decades back to the history. It is true, this tradition was broken for substantial time in seventies and eighties of the past century, but there was a solid fundament to build up on. This, together with technical and technological skill of the people in the country, helped significantly when re-establishing activities in this field two decades earlier. And the same aspect hopefully helps also in the future development.

3.1.1.1 Sixties of 20th century and alkaline systems

First systematic research in the field of hydrogen technologies, or more specifically fuel cells, was related to the nickel and KOH-based systems in sixties of 20th century. Possibility to utilize this technology as a power source was studied in the development laboratories of ČKD Praha, Semiconductor Division. They succeeded in the construction of 1 kW stack. This was a logical choice due to several reasons. As the main ones the following may be given:

- Alkaline system is relatively undemanding regarding the materials needed. It concerned primarily the polymer electrolyte required for the acidic PEM systems not accessible in the Soviet bloc in sufficient amount and quality.
- ČKD Praha's main focus was electric traction and alternatives to the traditional approaches were sought.
- Alkaline systems have still globally represented the main trend in this field at a given time period.

Unfortunately, this research was abandoned relatively soon in the beginning of eighties. Two-fold reasons for this may be considered. One of them is represented by limited resources available especially for the fuel cells components to be imported from outside Soviet bloc. Another aspect represents the fact that technology development under this political ideology was often limited to the optimization of the existing approaches over the introducing innovative, break-through technologies. Interesting is also the fact that the research was stopped almost at the same time as it was becoming very active in the technologically developed countries due to the oil crisis. Also, in this case the reason was political. In order to strengthen its political coalition, Soviet Union was keeping prices of the raw materials, including oil, constant and independent of the world market situation. This prevented the need for rapid technological development targeted to reduce dependence on the energy sources and push to the energy-saving approaches, including hydrogen technologies. It was also the reason why fuel cell technology was out of the scope of the Czech researchers and industry for almost 30 years. Unfortunately, some traces of this policy remain living in the basic approaches of current political representation to the technological development and to the future securing society by energy.

3.1.1.2 Nineties and revitalization of the interest

Back to the focus moved hydrogen technologies gradually in the middle of nineties of 20th century. Several aspects stand behind this development:

- Loss of traditional domains of research at mainly technical universities connected with crash of significant part of the Czech economy. This situation was connected with restructuralization of industry and overall economy of the country after the communistic period end.
- Search for new domains of research allowing building up on the existing competences, and at the same time being competitive in a broader international context.

- Changes in the financing of the science and research becoming from institutional significantly more, or even predominant, project oriented. This required establishing new, modern research topics.
- Strong need to establish new international relations allowing to solve more complex research problems and to gain access to the larger research infrastructures not available in The Czech Republic at that time.

From these aspects clearly follows that the lead lied mainly on the universities and research institutions. To this fact corresponded also problems studied and the structure of the research outcomes being concentrated mainly in the scientific papers. Industrial development, and thus practical applications, stayed within this period mostly out of the scope, with few exceptions to be discussed later on. Nevertheless, this pioneering time was, and still is, important from several aspects. It re-established this technological field in The Czech Republic and introduced back its basic research and development infrastructure. At the same time, it started to disseminate basic information not only within the scientific community but also among public. At the same time, it prepared background for the shift to the next step, that is, establishing hydrogen technologies as an independent specific field of not only research and development but also of high-tech industry.

A number of subjects have participated in this stage of development of the hydrogen technologies in The Czech Republic; selected ones of them, together with basic description of their role, are introduced in the following:

- **University of Chemistry and Technology Prague (UCTP)**

UCTP, a public university, is one of the subjects standing behind revitalising interest in the hydrogen technologies in The Czech Republic. It became active in this field in the second half of nineties of the previous century. First activities targeted development and testing of new, alternative polymer electrolytes. This research was performed in collaboration with the Institute of Macromolecular Chemistry of the Czech Academy of Sciences (IMC CAS). After low-temperature PEM fuel cells, high-temperature PEM fuel cells have been addressed, followed by the membrane water electrolysis. In the past years, a field of high-temperature solid oxide processes has been opened. In parallel also topics of hydrogen purification and storage have been studied.

Besides collaboration with IMC CAS, there were other two aspects boosting significantly these activities at UCTP: (i) intensive collaboration with ÚJV Řež, a.s. on demonstration and development activities and (ii) participation in a series of EU funded collaborative projects. Both activities provided necessary funding and contacts helping in rapid learning and establishment of this research and development field.

In the final step mathematical modeling activities allowed deeper understanding of the phenomena observed experimentally and their transfer to practical application. It also allowed intensive involvement in a broader spectrum of the research activities, especially on an international level.

It has to be mentioned at this point that UCTP accredited successfully study program on both bachelor and master level specialized on hydrogen and membrane processes. First students of this program are going to graduate within 2017.

- **Institute of Macromolecular Chemistry of the Academy of Sciences of The Czech Republic (IMC CAS)**

IMC CAS, a public research establishment, plays an important role in the hydrogen-related research activities in The Czech Republic. It possesses unique expertise in the field of polymer electrolytes development and synthesis. As such it participates in the numerous national, as well as international research activities. The main strength consists in combination of their effort with partners allowing for testing produced novel materials and partners allowing their introducing into the technical applications.

After initial focus on the cation selective materials as alternatives to the perfluorinated sulfonated acids of the Nafion type, within past years their activity targets mainly the development of a stable and sufficiently conductive alkaline polymer electrolytes allowing effective long-term application in the alkaline energy conversion systems. They become increasingly attractive during recent decade due to their significantly limited requirements on the cell and especially electrodes construction materials.

- **J. Heyrovský Institute of Physical Chemistry of the Academy of Sciences of The Czech Republic**

The J. Heyrovský Institute of Physical Chemistry of the Academy of Sciences of The Czech Republic is successor of the Institute of Polarography. Thus electrochemistry has long tradition there. Besides batteries the focus was paid also to the fuel cell technology. The non-platinum catalysts for alkaline fuel cells were studied at the Institute of Polarography in 1960s. Later on, their focus turns more to the batteries. The interest of research groups in the hydrogen technologies grows up again in past 15 years. Mainly the basic research of PEM system components is realized there.

- **Czech Technical University in Prague**

The Czech Technical University in Prague is one of the biggest and oldest technical universities in Europe. The hydrogen technologies fall primarily to the scope of energetics and transportation domains here. To get basic knowledge of PEM fuel cell characteristic, the demonstration laboratory equipped with two 1 kW commercial units was built in 2004–2005 on the Faculty of Electrical Engineering. Currently, more active is the Faculty of Mechanical Engineering focusing to the hydrogen as fuel for transportation.

- **Technical University of Liberec**

In agreement with the tradition of its research activities, an issue of engine with internal combustion using hydrogen as a fuel was studied here. The activities started already in 1997. Fuel system for direct injection of hydrogen to the engine valves was developed.

- **VŠB—Technical University of Ostrava**

The team in Ostrava focused from the beginning to the implementation of PEM fuel cell stack into the energy-efficient vehicle designed for Shell Eco-marathon. The first vehicle construction Hydrogenix started in 2004 on the Faculty of Electrical Engineering and Computer Science and continues up to now. Later in 2008, independent stand-alone energy system with possibility of accumulation of electricity produced by PV to hydrogen by PEM electrolyzer was developed and realized.

- **ÚJV Řež, a.s.**

ÚJV, a commercial engineering company, focuses on the problems connected with energetics in general. As such, it established in the first decade of this century also a group focusing on the problems related to the application of hydrogen in energetics. The topic of the first important project was the development of novel bus for public transportation powered by the low temperature PEM FC. After design of the new type of bus characterized by a triple hybrid power supply (fuel cells, batteries, and super capacitors), prototype production a homologation followed. But this topic will be discussed in a more detail later on in this text and in the Chapter 4 of this book. ÚJV also participated in the project ZEMSHIP dealing with development and implementation of the fuel cell powered ship in Hamburg.

It has to be stressed here that UJV represents an important part of the Czech community activities in the field of hydrogen technologies. Its main strength consists in its ability to implement recently obtained knowledge to the demonstration activities or wider technical applications.

- **Centrum Výzkumu Řež s.r.o. (CVR)**

Centrum Výzkumu Řež s.r.o. (Research Center Rez) was founded in 2002 as a daughter company of ÚJV Řež, a.s. Its central activities are oriented into the field of nuclear reactors and high temperature processes. Therefore, high-temperature fuel cells and water electrolysis are part of the research topic here.

- **Faculty of Mathematics and Physics, Charles University in Prague (FMP CU)**

FMP CU, a public university, has primarily focused on fundamental type research oriented toward electrocatalysis in the low-temperature PEM FC. A novel electrocatalyst material based on the utilization of Pt on a CeO_2 support was designed. It is characterized by the extremely high Pt utilization and thus low precious metal loads. Later on the activities were extended toward design and production of the commercial low temperature PEM FC test stations.

- **Astris, s.r.o.**

The Astris company was founded in 1992 in Benešov (The Czech Republic) as a daughter company of the ASTRIS ENERGI based in Mississauga (Toronto, Canada). Its main focus was originally on development of the electrodes for alkaline fuel cells. But the focus was rapidly widened on fuel cell systems from sub-kilowatt to 10 kilowatt level. As an outcome a stationary 4 kW generator and the portable 500 W generator models were produced.

At the onset of the new century ASTRIS embarked on an ambitious project, construction of a semi-automatic pilot production line for fuel cell electrodes. The company acquired a larger building and moved there in 2003. A new product, the POWERSTACK^{TM} MC250 was introduced, with significantly higher power density over the earlier LABCELL^{TM} stacks. The POWERSTACK^{TM} product then became the heart of the model E7 generator, a 36 V 1.8 kW engine fitting under the seat of a golf car, compact and more powerful than its predecessor E6, and of the E8 portable generator, most popular in the 48 V 2.4 kW version, suited to the use in telecommunications systems.

In this decade also new type of electrodes was produced and tested. For the first time the electrodes without any noble metal catalyst were produced and applied, a unique achievement in the field of low temperature, instantly starting fuel cells. Another important aspect of this work was modifying the manufacturing process to make it better adaptable to the future fully automatic mass production.

The ASTRIS product, a marriage of the fuel cell stacks manufactured in the Czech facility, and the balance-of-plant engineered and integrated by the parent company in Canada, was shown around the world. The E8 generator caught the eye of a multinational company, ACME Global based in India, looking for backup power supplies for its line of power sources for the mobile telephone transmitting towers. ACME acquired ASTRIS in 2007. The new owners transferred the entire manufacture of the E8 generator to Vlašim (The Czech Republic), where this work continued until 2010. After that the activities of Astris in Czech gradually diminished.

3.1.1.3 Economic and Political Background of The Czech Republic

Before going deeper to the questions related to the current situation and outlooks in the field of hydrogen technologies, a short view back to the economic and political background of the past 2 to 3 decades is important. It helps to understand some of the aspects and substantiations of the current development used later on.

As it was briefly mentioned earlier, existence of The Czech Republic in a Soviet bloc system for more than 40 years had a detrimental effect on the economy of the country. Very hard was also the impact on its political culture and on the ability to reflect flexibly visionary targets. These aspects become extremely visible after the system change in nineties of the 20th century. National economy was based, due to the communistic ideology, on the centralized management structure with minimum investment into the development and new technologies. It has clearly proven to be uncompetitive very shortly after opening the market. This led to the collapse and bankruptcy of major players, or, in a better case, to their privatization by foreign companies. In parallel to this, divisions of research and development have been typically closed. For almost a decade, applied research and an industrial development have been isolated to a certain degree from the industrial practice. Large companies owned by foreign subjects realized their research activities predominantly in the home country. Local branches were responsible mainly for the routine production using well-established processes and technologies.

Logically, different situation has to be considered for the local industry, that is, enterprises owned and controlled by the locals. Here, however, it has to be considered that a most of these enterprises has started from the small establishments involving few units of employees. Their competitiveness was built up on the extremely high flexibility and operability. But their resources did, and in fact still do not allow even partly strategic research to be conducted. Return of investment into the development has to to be relatively rapid.

Thus, their ability to react on the foreseen needs of a human society, and thus on future potential market, is strongly limitted. Although surprisingly many of these companies have achieved successful growth during the past two decades, their economic strength is still not high enough to represent fully competitive partners in such competition to the large international concerns. Fortunately, their position is improving gradually. Nevertheless, additional time is needed to reach sufficient financial strength. Here, a role of central government and of its policy supporting strategic development is crucial.

The main point in this represents the fact that a generally accepted long-term vision is necessary. To be stable it has to be accepted across the political spectrum, as well as by the majority of population of the country. Only this approach is providing a necessary background and allows the industry to open new strategic development pathways leading to the desired applications. It concerns into the certain degree financial support from the public funds allowing especially in the first phase of the high-risk research to keep the process economically viable. But maybe more importantly, it provides the basis of a suitable market as soon the technology has reached readiness level for commercialization.

Unfortunately, it seems that the common trend in the so called new democracies, including The Czech Republic, is the absence of such long-term vision. The reasons are manifold. One of those important seems to be high polarization of the political spectrum. It makes agreement across the political representation very difficult. Second issue represents expectation of rapid profit from any decision made. In politics, it means clear visibility of the results within the framework of the elections period. Unfortunately, this is not realistic in the case of strategic technology development. Important is also underestimation of communicating the visions to the electorate and to the public in general. More straight forward topics are looked for. In this sense, significantly more effort is needed in dissemination of hydrogen technologies to the public and thus in initiating the bottom-up approach. It is clearly very difficult and long-term approach, but it currently seems to be the only way leading to acceptance of the hydrogen technologies as a serious option in The Czech Republic. An important hope represents also development of the national industry and its gradual strengthening.

3.1.1.4 Role of the EU and its Strategy

The European Union has played a significant role in the development on the Czech national scene, although public has a tendency to simplify its influence, and at the same time to limit it, just to the regulatory aspects. But its role is from the point of view of authors of this text, significantly more complex, and it deserves few more words to understand it.

3.1.1.4.1 *Regulatory Role*

This role was mentioned few rows earlier as the only one people have typically on their mind. It is correct that the regulatory role played and plays an important role also in the case of hydrogen technologies; even it has not to be overestimated. Up to now any direct regulatory action directed specifically toward hydrogen technologies was in fact omitted. But as an activity targeted, between others, also to this field, there is the pressure to limit the environmental pollutions (especially CO_2, but also the others) and to strengthen the role of renewables in the energy mix and in the transportation. Hydrogen is mainly supported as it offers solution to several challenges connected with this strategic target. Within this framework The Czech Republic was also pushed to start to consider modern trends in the securing energy and in the transportation, although this sector was significantly discredited in the past. This was due by an unsatisfactorily managed public support of energy originated from photovoltaics. It was followed by negative medialization of all renewables as discussed in the following. The topic now gradually returns into considerations and, between others, thanks to the pressure on the CO_2 emissions reduction it is getting more and more discussed in connection with public support. This trend will hopefully include also hydrogen technologies. This is the first prerequisite of supporting implementation new technologies into the daily practice, for example, by corresponding regulations of the environmentally less friendly approaches.

3.1.1.4.2 *Inspiratory Role*

Besides regulations, there is also another approach in bringing hydrogen technologies closer to the market EU in using. We may mention here two of them playing significant role in The Czech Republic right now.

1. EU funded projects

Research and development projects funded by European Commission represent a valuable tool in motivating development of the hydrogen technologies within The Czech Republic. The impact of these projects is multiple. At first, it motivates intensive international exchange of knowledge and building of the international networks. The important aspect represents interconnection of the research and academic institutions with commercial or industrial subjects.

The research work performed is thus focused on the rapid progress in the technologies development and in rapid transfer of the know-how from research to the industry. Last, but not least, it allows to complete complex projects hardly realizable on the national level. This was a motivation of Czech subjects to participate in this type of projects dealing with the issue of hydrogen technologies since the 5th framework program of EU, that is, even before entering EU as a member country. Participation has helped significantly in improving level of knowledge and competences in this field. This is not easy to be quantified, as until 6th framework program (including), research activities were distributed among different types of calls and programs.

The fact that European administration considers hydrogen technologies as a tool for emissions reduction and for increase independence of Europa on the fossil fuels seriously is documented by establishing public-private-partnership in this sector represented by the Fuel Cells and Hydrogen Joint Undertaking (FCH JU). The main aim is to speed up the commercialization of the research activities by action of related industry having an important word in the management of this organization. Selected Czech subjects followed this trend. As first one, ÚJV Řež, a.s. joined FCH JU followed by UCTP. Thanks to the financial support by Ministry of Education, Youth and Sports of The Czech Republic, it was possible to participate in the projects financed by this organization from its beginning. This extremely supported development within the sector in The Czech Republic. Gradually, different Czech organizations started to participate in FCH JU projects, although, up to now, no other one has joined the FCH JU as a member.

2. Structural funds

The final aspect discussed here is related to the implementation of the European Structural Funds (ESFs). This is connected with a unique opportunity of equilibrating level in the different regions of the Europa. Outside the city of Prague, significant resources were invested, between others, into the development of research infrastructure. In consequence, several infrastructures have been realized targeting, between others, also hydrogen technologies. As concrete examples the following new installations may be mentioned: New Technologies Research Center Pilsen, CVR, Center for Research and Utilization of Renewable Energy at Brno University of Technology, ENET Center Ostrava, and so on.

New Technologies Research Center Pilsen was established by the University of West Bohemia in 2011. Their activities consist mainly in the evaluation of physical properties of polymeric membranes for PEM fuel cells and thermodynamic analysis of the fuel cells.

At the CVR the Sustainable Energy Project (SUStainableENergy, SUSEN) was implemented as a regional R&D center in 2015. A robust infrastructure for sustainable R&D activities to support Czech participation on European effort for safe and efficient energy generation in Europe in the 21st century was established. Particularly research activities for generation IV reactors, fusion, safe operation of generation III and II reactors is strengthened, and research and development in hydrogen technologies with particular focus on the high temperature solid oxides system were started.

Center for Research and Utilization of Renewable Energy was established by Brno University of Technology in 2010. The Laboratory of chemical energy sources deals in the field of fuel cells predominantly with catalysts and electrodes for alkaline and PEM units.

Energy Units for Utilization of Non-Traditional Energy Sources (ENET) Center was established by VŠB—Technical University of Ostrava in 2010. Hydrogen Technology Laboratory's main objective is testing and verification of the operational characteristics of new technologies for the storage and distribution of hydrogen including island systems and techno-economic analysis.

Although this was often significant investment to establish new technology on the sites without corresponding tradition and background instead of developing and upgrading existing infrastructure, it definitely contributed to the increase of capacity of installed research infrastructure. After necessary grow-up period, it will help to cover extend of the research necessary to keep competitiveness and possible penetration of these technologies into the industrial practice.

The ESFs play also direct role in the implementation of hydrogen technologies in The Czech Republic. In the calls for projects planned within the framework of the forthcomming ESF program period, several topics focused on demonstration activities within the field of hydrogen technologies are foreseen. This could help in strengthening technological development, as well as public awareness. Initiation of extension of existing infrastructure then increases motivation to implement broader spectrum of technologies.

EU thus plays an important role in the endeavour to implement hydrogen technologies also on the nation level although often indirect.

It mainly consists in supporting activities developed by the locals. But any support in reaching target is highly welcome.

3.1.2 Czech Hydrogen Technology Platform and National Context

Most of the aforementioned activities have proceeded on an individual, or, in the best case, on a bilateral basis. Effective coordination of the activities, as well as an effective lobbying and public relation activities were almost completely missing. This had a negative impact on the visibility of the hydrogen technologies and on their success when competing for a broader support, both in a grant project competitions, as well as in penetrating in the broad awareness. This is, at least to a certain degree, connected with a specific situation of The Czech Republic as a post-Soviet bloc country discussed earlier. It does also influence discussion on a long-term policy in energetics and related fields of The Czech Republic.

3.1.2.1 The Czech Republic and Role of Hydrogen in an Energetic Policy

In The Czech Republic the energy mix is based on the traditional sources, mainly on the brown coal (49.7%) and nuclear (19.6%). Hydropower also plays an important role reaching up to 10.3%. Close to this number is also photovoltaics (9.3%), which is due to the subsidy policy of the Czech government in the years 2008–2010. Subsidy of the electricity originated from photovoltaics was set on an incorrectly high level, which motivated rapid growth of photovoltaic power stations of MW sizes. The following demands on the subsidy significantly impacted price of energy. This aspect was subsequently broadly medialized. As a consequence very strong negative publicity of all renewables arose in the Czech. This damaged public image of all renewables being considered as a way to increase public subsidy for private entities and is still causing serious problems. Any promotion of the modern approaches to the energy production independent from fossil or nuclear fuels is primarily considered as another trial of deception. Change of this situation requires significant time and effort.

Strategy in energy politics of The Czech Republic was not very clear for a long time. Therefore, there was established so called Paces Commission. Vaclav Paces, a former chairman of the Czech Academy of Sciences, was chosen as a chair of the Commission because of his high reputation and independence of the main lobbying groups to prepare an outlook to further development in The Czech Republic and to propose the most promising and safe ways of securing energy demands of the country for the next decades. The main conclusion made was to strengthen for the intermediate period nuclear energy as a technology replacing partly brown coal power plants, allowing thus to reduce CO_2 emissions. Development of renewables was not set as priority. This was possible because of exceptionally high acceptance of the nuclear energy by the Czech population. An important role in this position of the Czech public played negative campaign of the Austrian government and citizens associations focused against the start of operation of the Temelin nuclear power station. The negative style of these campaigns had an opposite impact on the national feelings of the Czech people. The second aspect is long-term publicity in the direction of effective utilization of the nuclear energy in combination of inability of the ecologically oriented initiatives to provide effective campaign against such policy.

This situation, of course, hinders rapid development of alternative approaches considered often, as already said, to be inadequately expensive and insufficiently reliable. This makes it rather complicate to develop alternative methods of energy production, conversion and storage, including hydrogen technologies, beyond the boarder of an academic research. For a long period of time, it was difficult to identify private subject willing to invest resources in the technologies beyond commercially widespread technologies. It is a matter of past few years when a barrier in this direction seems to "leak" gradually and to allow penetrating first information to the public. But it is a result of a long-term and intensive coordinated action of several subjects active in this field coordinated by the Czech Hydrogen Technology Platform.

The next two chapters analyse the impact of the Czech Hydrogen Technology Platform. This seems to be important to the authors as in other countries, for example, the neighbouring Germany, see also chapter "Germany" in this book, a different development at a very different pace is visible. The varying initial points are the reasons for this mosaic of different levels of hydrogen technology implementation, which is visible in not only in Europe but also in the World. And this needs to be taken into account for realistic estimation of achievable changes. Therefore, it is the object of these chapters to increase the mutual understanding in Europe and maybe also provide information for regions, which have even a similar or longer path to implementation than The Czech Republic.

3.1.2.2 Czech Hydrogen Technology Platform and its Impact

Czech Hydrogen Technology Platform (HYTEP) opened new period of hydrogen technologies in The Czech Republic. This organization was established in 2006 under auspice of the Ministry of Industry and Trade of The Czech Republic. Leading role in this activity has UJV Řež, a.s. and it has played also a leading role in running HYTEP for a number of years. From its beginning the HYTEP ambition was to coordinate activities of the individual subjects active within The Czech Republic in research development and deployment of the hydrogen technologies. This is a common statement given in the preamble of all platforms and associations of this type. Their position and main mission, however, often significantly differs. As one of the main reasons strength of the industrial sector represented by this association in the national or broader context may be given. In the case of well-established industrial sector with significant economic power, voice of unifying platform or association is significantly stronger and its impact clearly visible. Also public accepts such discussion significantly better, because they are aware of the significance and contribution of the sector to the society.

In newly establishing, not yet developed and commercially not significantly active domains the situation is different. In the case strong national industry exists having a long-term strategy in the given domain, it may play a role of an integrator. In absence of an important industry, or of a dominating partner, role of relevant association is important, as it allows unifying power of the individual players and optimising of using resources in defending the technology and promoting its advancement. The role of HYTEP was and still is of exactly this type. It unifies the area of hydrogen technologies scene, which is currently not clearly dominated by any commercial body in The Czech Republic. It gradually developed, thanks to the activity of the members, to a common platform helping them to make their voice hearable. This is especially important aspect when discussing with public bodies.

After stabilising internal situation, HYTEP set its long-, intermediate- as well as short-term targets. Strategic long-term target is to integrate HYTEP to the European and broader structures as a strong partner able to protect and promote interests of its members. From the intermediate point of view, increase in a number of commercial members and establishing support of hydrogen technologies as a specific part of the national research agenda, represent the most important factors. An important activity is represented also by establishing the Central and Eastern European network helping to overcome handicap of these countries caused by the "post-communistic heritage", that is, lack of the national strong industry with long-term vision and economic strength allowing performing strategic research and development discussed earlier. Short-term target is then strong and clearly targeted information campaign making hydrogen technologies a widely known topic understandable to the public, as well as to the decision makers. An important aspect represents also promotion of hydrogen technologies between commercial subjects. In this case is the target to motivate commercial subjects to become involved in this exciting field. Initiation of the internationalization process represents another aspect.

It is thus clear that in the so called "new member countries," the role of the associations or technological platforms is potentially very important if their management is adequately active and resources for their operation are available. Also in this case, however, relatively long-term stability, continuity, and vision are necessary to achieve difficult targets set.

3.1.2.3 HYTEP Successes and Lessons Learnt

How successful is HYTEP so far in addressing its targets and in promoting hydrogen technologies on the local, as well as on the broader scale? Certainly, several fails were encountered in the history of HYTEP counting already 10 years. Probably the most important one is connected to variation in the number of member subjects, especially from the commercial sphere. This was connected primarily with unrealistic expectations of the members joining HYTEP in the early stages after its founding. Expectations related to the funding strategy of the Czech government and fast monetization of the existing activities represented the main motivation aspects at this period of HYTEP operation. As soon as these expectations become clearly too optimistic, structure of the membership has changed significantly. Since that the restructuralization is going on providing membership structure possessing more realistic expectations and targets connected with their activities in the field. It is also connected with very promising development observed within past years. HYTEP become significantly more active in this period.

Three main successes may be identified now. Stabilization of the membership structure and increase of number of the commercial subjects count as the first one. It is connected with the second success represented by the gradual penetration of the knowledge of hydrogen technologies to the industrial sphere as well as into the state and municipalities administrations. The third success is related to the increase in the visibility of HYTEP in an international context.

The participation of the commercial subjects in the HYTEP activities represents an important aspect from several points of view. This is clearly connected with the final target of HYTEP endeavour, that is, transfer of the hydrogen technologies from the domain of the research and development toward a practical application. Of course, as a first step demonstration activities are necessary allowing evaluating operational characteristics of the selected technologies and their performance under real conditions. On the other hand, also collection of necessary experience from operating such systems is a necessary prerequisite on the road to broad utilization of any new technology. Without participation of the commercial subjects, such target is not reachable. Research establishments do typically not possess corresponding expertise and technological as well as economical background to perform such tests, especially on a sufficiently long time scale. Moreover, commercial subjects have significantly more advanced tools and processes to evaluate commercial potential of different options under scope. Only this approach allows to choose the most promising alternatives. Therefore, an increasing interest of these subjects represents an important and positive signal toward future development.

Increased visibility of the hydrogen technologies is a consequence of an intensive campaign organized continuously by HYTEP. It is connected with an attendance of its members at different seminaries and workshops organized by commercial subjects as well as by the public administration. Regular visits at responsible representatives of various public institutions, ministries, and so on play a significant role. This is a necessary prerequisite to the opening of basic communication channels and thus to establishing first activities and rising first interest. First communication in both directions, that is dialogue, offers a good opportunity for both sides to clarify the situation to see the benefits and challenges connected with a progress in the direction of implementing of the hydrogen technologies. Without such analysis a targeted support from public sources and involvement of hydrogen technologies in the research programs of the governments is only hardly possible. At the same time, support of the municipalities is necessary in order to be able to collaborate with them on the selected demonstration activities providing potential solution to the problems to be solved (e.g., local emissions).

International visibility, and thus an increasing potential for the broader collaboration and coordination, represents an important aspect at high-tech technologies in the phase of rapid development and first introduction to the commercial market. HYTEP has chosen in this respect to organize an international conference Hydrogen Days. Although tradition is going back for almost entire decade of HYTEP existence, the edition organized in the year 2014 represented the break through. For the first time the event was broadly promoted on an international level. A high number of participants from a number of countries attended the event and reported on their research, development, and implementation activities. A number of registered attendees has reached 51 people originating from 6 countries. This has created a very good base for the following years. The edition 2015 focused on creation of new strategy and its promotion. It was for the first time countries of the former Soviet bloc were individually motivated to join. Although only partly attended by their representatives, a tradition was created. It allowed more effective promotion of continuation of this endeavour. The 2015 edition was characterized by 50 registered participants, but already from 11 countries, that is, almost doubled with respect to the previous year. On the base of the experience gained, new edition 2016 was organized with a support of the Visegrad Funds. Focus was put on the networking between the Post-Soviet bloc countries and their collaboration with countries advanced in implementation of the hydrogen technologies. This process is beneficiary for both sides. Joining of the east Europa countries makes their "voice" stronger. At the same time it prepares market opening for the hydrogen technologies in this region. Without activities of the local organizations and commercial subjects such process will be strongly delayed and connected with a number of obstacles. Therefore, it is in an interest of both sides to further promote this process. The attendees' structure clearly documents gradual internationalization of the event. At this year 85 registered participants arrived from 15 countries. Additionally 30 guests participated, predominantly from the ministries of the Czech government, municipalities and media. Additionally, this event was for the first time broadly supported by the public as well as scientific sphere. Its auspice have provided Deputy Prime Minister for the Science, Research and Innovation of The Czech Republic, Ministry of Industry and Trade of The Czech Republic, Ministry of Transport of The Czech Republic, Ministry of the Environment of The Czech Republic, Ministry of Education, Youth and Sports of The Czech Republic, Municipal District of Prague 6, and Municipal District of Prague 7. The support of the International Society of Electrochemistry was also highly welcome.

Interesting information connected with recent development in this direction represents a fact that in the year 2017, World Hydrogen Technology Convention, a traditional event of International Association for Hydrogen Energy, is organized in Prague. For the first time it is organized in an East

European country clearly documenting gradual penetration in this direction and increasing acceptance of the countries from this region in a broader international structures.

3.1.3 Current Activities of the Major Subjects in the Field of Hydrogen Technologies in The Czech Republic

This chapter has no ambition to provide a complete overview of all activities in the field of hydrogen technologies going on in The Czech Republic. It represents a broad field with a different degree and way of involvement. Besides basic research on particular problems realized by one group of subjects, more complex approaches ranging from the basic research to the focus on the aspects related to the transfer of the technology to the demonstration or real life units are visible at the second group of institutions. Subjects focusing on the knowledge transfer and technologies verification on a pilot, or industrial scale, represent a final important step.

3.1.3.1 Fundamental research activities

Fundamental research activities address typically the issues of electrocatalysis, electrode properties, and solid electrolytes development. Mainly universities and institutes of the Czech Academy of Sciences are involved in this part of the activities spectrum. Fields of interest followed by the individual institutions may be in brief characterized as follows:

- *UCTP*

Development of novel catalysts for the water electrolysis, optimization of the electrodes and MEA construction, degradation processes analysis, and mathematical modeling are the main fields of interest. At the same time it collaborates on a number of activities developed in the field of basic research by IMC AS CR.

- *IMC AS CR*

Development of novel polymer electrolytes, both in a form of polymer electrolyte membrane, as well as in a form of catalytic layer binder, recently with main focus on alkaline systems.

- *FMP CU*

Development of low Pt content catalysts, study of principles of their function by means of highly advanced spectroscopic techniques.

3.1.3.2 Development and knowledge transfer activities

This part of process of application of the research results represents probably the most difficult one. It is time and resources demanding, while providing no rapid, well visible results, no guarantee of final success. It is even difficult to produce successful scientific publications, an important outcome for the academic bodies. Throughout all these facts, this step is an inevitable part of the implementation of new technologies. As such it has to be treated carefully and in all details. Typically, close collaboration of universities and other academic bodies with an engineering company provides best chances for the success.

We come across similar situation also in the case of hydrogen technologies here in the Czech. The transfer activities are in this case usually initiated by ÚJV Řež, a.s., an engineering company. Collaboration with academic institutions is supporting these activities as described. UCTP represents the partner most intensively involved in this collaboration. The main activities have been focused recently on the problem of high temperature (solid oxides) steam electrolysis and alkaline water electrolysis with polymer electrolyte membrane. These problems are closely related to the current direction of development, that is, evaluating possibilities of the intermittent storage of electrical energy originating predominantly from photovoltaics and alternatively from the wind and from the helium cooled nuclear reactors of the fourth generation. The target for the future is to extend the activities toward cogeneration based on the solid oxides fuel cell systems. This technology offers significant application potential in a scheme of distributed energy supply, especially in the domain of relatively small production capacities ranging up to 100 kW. A test unit is set up at the CVR to investigate the hydrogen production in connection with high-temperature processes such as generation IV reactors.

3.1.3.3 Demonstration activities

ÚJV Řež, a.s., in collaboration with commercial subjects focused on hydrogen technologies and broader concepts are mainly active in this field. These subejcts are also related to the two main demosntration activities realised in The Czech Republic so far.

TriHyBus (triple Hybrid Bus) was a result of the project realized in the years 2006–2009. The public transportation bus was constructed using electric propulsion with energy provided by a unique system consisting of PEM FC stack, batteries, and ultracapacitors. It was first hydrogen-powered bus in former Soviet bloc countries. Currently, it is still operated by Veolia Transport under the management of ÚJV Řež, a.s.

Hydrogen filling station was put in the place as a follow-up of the TriHyBus project. It is located in the Veolia Transport bus park in Neratovice, city near Prague. The filing station is currently using hydrogen pressure of 350 bars, but is designed and can easily be upgraded to 700 bars.

Energy storage system based on 13 kWp photovoltaic power plant, PEM water electrolysis, hydrogen container for 10 kg H_2, and PEM fuel cell was constructed directly in the premises of ÚJV Řež, a.s. in 2013. The aim is to optimize the energy management. Similar but significantly smaller system is also installed in the research center ENET.

3.1.4 Challenges and Promises

In the past an interesting phenomenon connected with the hydrogen technologies consisted in the fact that the interest in their development and implementation occurred in cycles. They have usually been connected with the oil crisis or another external factors rising fear of lack of corresponding energy resources, especially in transportation, although activities in the stationary systems were originally in a scope. Within the past decade, situation started to change gradually. This is connected with relative matureness of the technology for the mobile applications and with the increasing emphasis put on the renewable energy sources. As already discussed, it is connected with a strong need of sufficiently flexible and efficient systems able to store local (in time and position) excesses of the energy produced. Since installation of the renewable energy sources represents a continuous process, which has already reached a significant portion of energy production/consumption capacities, this interest can be considered for continuous and gradually growing. So, the cyclic characteristic is in this respect, hopefully, overcome.

This expectation may be supported by several facts. The first one represents an increasing number of private enterprises being active in the field accompanied by an increasing range of commercial products available on the market. Unfortunately, this movement differs in its strength across the Europa. Reasons are indicated earlier. Whereas economically stronger and technologically more developed countries clearly pay an increasing attention to this segment of energetics, second group is ignoring this trend, or, in a better case, waits for the outcome of the ongoing development before joining it. European Commission is aware of this trend and is trying to provide sufficient background for the as homogeneous as possible development of this technology among the member states and for strengthening Europa competitiveness in this segment of energetics. An increasing attention being paid to the low (or zero) emission processes helps this technology to penetrate deeper into the awareness of decision makers on a national level. Unfortunately, once again, this process is not happening equally rapid in all countries. This creates undesired asymmetry between the individual territories and thus endangers successful and rapid implementation of hydrogen technologies into the practical life.

Nevertheless, an important effect of the activities on a European level is significantly increasing awareness of not only decision makers but also of a general public. This is one of the key factors from several reasons. First one represents the fact that development and implementation of most of the hydrogen technologies is still supported from the public funds. So, the tax-payers awareness is an important aspect. Additional to this, only the informed public can establish the corresponding market for the new technologies. And increasing number of, at least potential, customers is the best motivation for the commercial subjects to enter this field. Unfortunately, also in this aspect Europa is still divided. Whereas in part of the countries number of installations and real-life applications has already reached an important number, in the other countries, especially the so called New European ones, situation is not so well developed. But progress in a positive direction is notable. Nevertheless, in the second group of the countries the situation is, due to the problematics being still under development, not very stable. In the case of problems occurrence, which will be more interesting for media and understandable by public, the implementation of this particular technology may be significantly delayed.

Of course, all the earlier discussions are realistic under condition of sufficient matureness of the hydrogen technologies. In a number of sectors this is already true. As a good example may serve mobility applications as buses or personal cars represented not only by the broad demonstration activities but also by first series production activities, for example, Toyota Mirai. But an increasing number of other technologies may be mentioned, for example, hydrogen filling stations, water electrolyzers, (micro)cogeneration systems and others. So, technological matureness shall not represent a major obstacle for a number of applications. It is thus necessary to search for the true reasons, as usually, in the economics. It is clearly visible that in the predominating number of potential applications hydrogen technologies are not yet economically competitive. Very often economy of scale is given as a main substantiation of this fact. And it is definitely one of the very important aspects. At the same time, however, the history of development and innovations has to be considered in this respect. Whereas competitive technologies are developed for a number of decades with an investment higher by several orders of magnitude, hydrogen technologies are intensively studied for the past two decades. To gain this lack will still require a years of investment. On the other hand, good signal is coming from the fact that in selected fields it is already viable and reaches wide spreading without additional support. The important question is thus in front of the decision makers. For rapid implementation and further development corresponding regulations in the field of emission limitations and increasing share of renewable energies are foreseen. This could boost the penetration of hydrogen technologies into the free market. In the case of no support an important danger for attained know-how exists. A significant part of currently running enterprises will with high probability stop their activities, as it would be difficult for them to finance long-lasting transition period from private sources. Side effect of this represents a loss of teams, experts, and their know-how. It would be very difficult to bring it again up on a comparable level. These aspects have to be considered carefully when establishing national, as well as Europeans strategies for energetics and mobility development.

As a final aspect people's expectations have to be mentioned on this place. Originally, informed people might be divided to the two groups: (i) enthusiastic and (ii) sceptic. No one of these positions does at the end really helps the technology to be rapidly implemented. Too enthusiastic position often does not allow to see the critical drawbacks and thus to focus on their solving, or minimising their impact. Sceptical position is in opposite situation. Due to the existing drawbacks does not see the real value of this approach. The motivation for further progress is thus missing. It is very positive signal that the problem has moved from the emotional to the rational level within the past years. Rational discussion helps to identify bottlenecks and to find corresponding solutions. This is the only way how to reach the desired level of implementation.

The authors hoped to show that The Czech Republic has to find its unique way to innovative approaches in energetics, like hydrogen technologies, in the triangle of its heritage as a former Soviet bloc country, the current European circumstances (legislation and innovation support) and its own strong tradition as an industrial country with a core area in heavy industries and a centralized infrastructure. This led to a specific situation with several obstacles. But an increasing interest in hydrogen technologies within the country has led to a growing awareness and the homemade problems are now tackled. To conclude, hydrogen and hydrogen technologies are on a good way to the broad implementation in a broad range of technologies. This period is, however, connected with significant risks. As typical examples costs related to establishing missing infrastructure, missing legal framework and public experience, as well as competition with the well-established technologies may be mentioned. Only sufficiently large market may help to solve these obstacles. A lot of attention is, therefore, necessary not only in the development of the technology but also in preparing corresponding environment for its introduction to the market.

References

[1] Bouzek, K., Holzhauser, P., Kodym, R., Moravcova, S., Paidar, M. Utilization of Nafion (R)/conducting polymer composite in the PEM type fuel cells. Journal of Applied Electrochemistry. 2007; **37**(1): 137–145.

[2] Bouzek, K., Paidar, M., Malis, J., Jakubec, I., Janik, L. Influence of hydrogen contamination by mercury on the lifetime of the PEM-type fuel cell. Electrochimica Acta. 2010; **56**(2): 889–895.

[3] CENTEM. (2016). "Research Program—Advanced Technologies Based on Polymer Materials", 2016, from http://ntc.zcu.cz/en/odbory/CENTEM/CTx/

[4] Center for Research and Utilization of Renewable Energy (CVVOZE). Chemical and photovoltaic energy. 2016, from http://www.cvvoze.cz/en/chemicke_a_fotovoltanicke/index

[5] Center of Vehicles for Sustainable Mobility (CVUM). Project Center of Vehicles for Sustainable Mobility. 2016, from http://www.cvum.eu/en/project-cvsm

[6] Chanda, D., Hnat, J., Paidar, M., Schauer, J., Bouzek, K. Synthesis and characterization of $NiFe_2O_4$ electrocatalyst for the hydrogen evolution reaction in alkaline water electrolysis using different polymer binders. Journal of Power Sources. 2015; **285**: 217–226.

[7] Commission, E. (2015). Operational Programme 'Research and Development for Innovations'. 2016, from http://ec.europa.eu/regional_policy/en/atlas/programmes/2007-2013/czech-republic/operational-programme-research-and-development-for-innovations.

[8] Eet CENTER. (2015). Hydrogen Technology Laboratory. 2016, from http://enet.vsb.cz/en/science-research/laboratory-wp03/hydrogen-technology-laboratory/.

[9] Hnat, J., Paidar, M., Schauer, J., Zitka, J., Bouzek, K. Polymer anion selective membranes for electrolytic splitting of water. Part I: stability of ion-exchange groups and impact of the polymer binder. Journal of Applied Electrochemistry. 2011; **41**(9): 1043–1052.

[10] Hnat, J., Paidar, M., Schauer, J., Zitka, J., Bouzek, K. Polymer anion-selective membranes for electrolytic splitting of water. Part II: Enhancement of ionic conductivity and performance under conditions of alkaline water electrolysis. Journal of Applied Electrochemistry. 2012; **42**(8): 545–554.

[11] Jindra, J., Heyrovský, M. The Czech (Czechoslovak) Electrochemistry 1900–1990. Electrochemistry in a Divided World: Innovations in Eastern Europe in the 20th Century. F. Scholz. Cham, Springer International Publishing: 13–48.

[12] International Energy Agency (2010). Energy Policies of IEA Countries—The Czech Republic. https://www.iea.org/countries/membercountries/czechre public/

[13] Mališ, J., Mazúr, P., Paidar, M., Bystron, T., Bouzek, K. Nafion 117 stability under conditions of PEM water electrolysis at elevated temperature and pressure. International Journal of Hydrogen Energy. 2016; **41**(4): 2177–2188.

[14] Mazur, P., Malis, J., Paidar, M., Schauer, J., Bouzek, K. Preparation of gas diffusion electrodes for high temperature PEM-type fuel cells. Desalination and Water Treatment. 2010; **14**(1–3): 101–105.

[15] Mazur, P., Polonsky, J., Paidar, M., Bouzek, K. Non-conductive TiO_2 as the anode catalyst support for PEM water electrolysis. International Journal of Hydrogen Energy. 2012; **37**(17): 12081–12088.

[16] Mazúr, P., Soukup, J., Paidar, M., Bouzek, K. Gas diffusion electrodes for high temperature PEM-type fuel cells: Role of a polymer binder and method of the catalyst layer deposition. Journal of Applied Electrochemistry. 2011; **41**(9): 1013–1019.

[17] Minařik, D. Experimentální projekt energetického systému s akumulací elektrické energie na bázi vodíkový ch technologií. Proceedings of the 12th International Scientific Conference Electric Power Engineering 2011, EPE 2011.

[18] Minařik, D., Horak, B., Moldrik, P, Slanina, Z. An experimental study of laboratory hybrid power system with the hydrogen technologies. Advances in Electrical and Electronic Engineering. 2014; **12**(5): 518–528.

[19] Mrha, J., Iliev, I., Kaisheva, A., Gamburzev, S., Musilova, M. On the effect of various active carbon catalysts on the behaviour of carbon gas-diffusion air electrodes: 2. Acid solutions. Journal of Power Sources. 1976; **1**(1): 47–56.

[20] Paidar, M., Vazac, K., Roubalik, M., Bouzek, K. Alkaline water electrolyzis with perfluorinated cation-selective membrane. Desalination and Water Treatment. 2015; **56**(12): 3203–3206.

[21] Polonský, J., Paidar, M., Bouzek, K. Influence of catalyst support and preparation method on properties of Pt catalysts for PEM type reactors. 19th International Congress of Chemical and Process Engineering, CHISA 2010 and 7th European Congress of Chemical Engineering, ECCE-7, Prague.

[22] Smrček, K. An investigation of the oxygen cathode with a silver catalyst polarized in the region of negative potentials. Journal of Power Sources. 1981; **7**(2): 105–112.

[23] Smrček, K., Beran, J., Jandera, J. The silver catalyst in the hydrophobic oxygen electrode. Journal of Power Sources. 1977; **2**(C): 273–286.

[24] Smrček, K., Marholova, O. Hydrogen-oxygen Fuel Cells. 1981; **70**(10): 574–577.

[25] Technical University Liberec. Hydrogen Engine with Internal Combustion. 2016, from https://www.tul.cz/en/research/sustainable-technology.

[26] TriHyBus. (2008). "The First Hydrogen Bus in New EU Member States." 2016, from http://trihybus.cz/homepage.

[27] Žitka, J., Bleha, M., Schauer, J., Galajdová, B., Paidar, M., Hnát, J., Bouzek, K. Ion exchange membranes based on vinylphosphonic acid-co-acrylonitrile copolymers for fuel cells. Desalination and Water Treatment. 2015; **56**(12): 3167–3173.

4

TriHyBus—An Unwanted Miracle in The Czech Republic

Jakub Slavík

Independent business consultant specialized in transportation and smart city projects, The Czech Republic

4.1 The Project

TriHyBus is an EU demonstration project led by ÚJV Řež. It is the only Czech and the first CEEC's fuel cell bus project and hydrogen filling station. The project including the filling station is located in Neratovice—an industrial town in the Mid-Bohemian region.

The project partners were

- ÚJV Řež (CR)—The project coordinator, safety and legislative issue, and the test operation;
- Škoda Electric (CR)—The electric propulsion, power management, and finalization of the bus;
- Proton Motor (Germany)—Development and production of the fuel cell, fuel tanks, and hydrogen infrastructure;
- Linde Gas (CR)—The hydrogen refueling station and hydrogen fuel supplier;
- IFE Halden (Norway)—Monitoring, information, and control technologies; MMI (Man Machine Interface) supplier;
- Arriva (earlier Veolia Transport)—The bus operator in Neratovice;
- The Ministry of Transport.

Important partners in the project preparation stage were also municipal and regional authorities concerned as well as TÜV SÜD Auto CZ as the vehicle roadworthiness certification body.

The project's total cost was € 3.3m, of this ca € 1m for the infrastructure and the remaining budget split about 50:50 for the bus development and for the bus manufacturing.

The project funding structure was as follows:

- ERDF: 56.25%;
- The Ministry of Transport: 18.75%;
- Own sources: 25%.

The project was implemented on the period between January 2008 and December 2009. The condition of the project sustainability was to keep the bus operation at least until the end of 2014.

The bus is still operable, but its future is unclear.

4.2 The Vehicle

ThiHyBus is a fuel cell bus prototype based on 12m Irisbus Citelis design with the overall capacity of 96 passengers, of which 26 only seated.

The powertrain: 120 kW Škoda Electric ML3444 K/4 asynchronous four-pole motor.

The triple hybrid design:

- 50 kW Proton Motor's PEM fuel cell unit containing 6 stacks at 100 cells each; the output voltage is 300–560 V and the max. current is150 A;
- 26 kWh Li-Ion batteries Valence Technology UEV-18XP;
- Four Maxwell's supercapacitors with the total capacity 1 kWh and power of 200 kW by acceleration and 300 kW by braking.

The vehicle is equipped by Škoda Electric hybrid energy management system and IFE Halden's man-machine interface system.

Clean hydrogen with 35 MPa pressure is used as the standard fuel by fuel cell buses. The hydrogen is stored in four composite tanks of 20 kg total capacity.

4.3 The Hydrogen Refueling Station

The hydrogen in the station is stored in a low-pressure tank (4 MPa). The amount is sufficient for a week operation of the bus. The compression station is located in a massive concrete container. For safety reasons, the interior space is physically divided into two parts. The first one contains control electronics, hydraulic compressor, and other electrical installations and the second one contains equipment for compression, transmission, and regulation of the flow of compressed hydrogen.

The filling station is supplied with hydrogen by Linde Gas weekly, depending on the actual needs of the project. The station was designed with regard to possible future increase of its storage capacity (in case of more hydrogen vehicles in operation) and an increase in boost pressure to 70 MPa.

4.4 The Operation

The TriHyBus sample operational results are shown in the Table 4.1.

The triple hybrid design brings very positive results, especially in the urban service in the rather flat landscape of Neratovice, where the bus has been tested.

During the operation, the fuel cell unit runs continually at 100% of its performance with no significant deviations. The regenerative braking stores the energy to the supercapacitors and the stored energy is immediately used by the acceleration. If not enough energy in the supercapacitors, the batteries get involved, too. When the bus reaches the necessary speed, the relatively weak fuel cell unit generates enough energy to keep it moving. When the bus is standing, the fuel cell recharges the supercapacitors and the batteries. If both of them were fully charged, the rest would be wasted in the resistors, as usual by electric vehicles.

As obvious, the batteries do not get much involved in the energy management during the standard "stop-and-go" urban service, but rather on longer

Table 4.1 Sample data from the operation

Observed period of time	Jan 2012 to Apr 2013.
Location	Neratovice urban transit, demonstration trips in various places.
Observed fleet mileage	1400 km of this ca 120 km in revenue service (the operator's estimation; exact data not available).
Average hydrogen consumption and price	7,75 kg/100 km—20 l diesel equivalent. Hydrogen price: CZK120 (currently ca €4.4) per kg.
Average range per tank single filling	275 km.
Tank filling time	10 minutes.
Other available operational data	Total availability including the gas filing station: 58% of this vehicle availability: 91% One vehicle failure occurred during the observed period: brake pedal electronics. Problems with the hydrogen refueling station availability—long waiting period for repairs.

straight sections. They are also used for the start-up process of the bus and the fuel cell system as well as for the backup energy storage.

4.5 The Project Story and Lessons

At the time of the TriHyBus project implementation, the hybrid fuel cell bus configuration was used, for example, by the NREL-funded fuel cell bus projects in the USA or by the European CUTE (hybrid bus prototype) and CHIC projects typically involved a ca 120–150 kW fuel cell unit as the main source of energy together with ca 11–50 kWh batteries for the power balancing. The average hydrogen consumption in this configuration usually ranged between 9 and 15 kg per 100 km.

The TriHyBus prototype vehicle price was very similar to fuel cell buses in the USA, as presented by the NREL.

In this comparison TriHyBus, using a fuel cell unit just about as powerful as the engine by a small passenger car and the rest of energy re-cycling due to extremely high power economy, looked nearly like a "perpetual motion machine". It is not surprising that in 2010, the project won the Golden Medal in the International Machinery Fair in Brno.

However, the project has had an inborn managerial problem that influenced its development and the current unclear future: the absence of a strong and fully committed project owner.

© Jakub Slavik

Information sources: NREL reports, EHA, CHIC, and TriHyBus project websites and the portal *www.proelektrotechniky.cz* operated by the author.

ÚJV Řež—the project co-ordinator—is a private institution involved mainly in nuclear and non-nuclear engineering. The TriHyBus project was an opportunity for its immediate presentation, but very obviously not a key stream in its corporate strategy.

For Škoda Electric, the project has been a very good opportunity for testing its energy management system with various power sources. However, its product portfolio for urban transit covers mainly powertrains for standard electric vehicles such as trams, trolley-buses, or metros.

While Proton Motor has used two generations of its fuel cell units in TriHyBus and has widely collected the data from its operation, it has not presented itself as a strong and committed driving force willing to invest in the project continuation.

As the operational data sample show, the project has been used mainly for demonstration purposes rather than for revenue service, which resulted in a rather low commitment and interest by the public transport operator.

At the moment, the key driving force for keeping the project somehow living is personal enthusiasm of some people involved in its implementation, such as the ÚJV Řež's project manager and chief engineer or the bus driver (actually the owner of the original local transport company acquired by Veolia), who make their best whenever "their" TriHyBus is concerned. It may be excellent by single opportunities, but this is not a strategy.

In the meantime, the world development by fuel cell buses has been moving fast forward.

There is a strong effort toward the fuel cell bus commercialization using the economy of scale pro-reducing the vehicle price. The target fuel cell bus price in the current FCH JU's tender is €650,000, that is, nearly half of the TriHyBus vehicle price (or manufacturing cost).

Heavy vehicle fuel cell unit output has been reducing, which results in reduced fuel consumption. For example, the new Van Hool fuel cell buses for Rotterdam use Ballard's 90 kW fuel cell unit and their anticipated consumption is ca 8 kg/100 km, that is, nearly as high as by TriHyBus. Instead of a rather complicated triple hybrid power system, the latest fuel cell buses use just batteries with enough operational flexibility for the power balance and fuel economy.

TriHyBus bus is currently at two thirds of a standard bus vehicle life. Instead of responding to these trends and testing new components and their configurations, it has been remaining a unique, ingenious, but single for ever prototype—not a pattern for the future production in series.

The project story shows a simple lesson: Even an ingenious technological idea brought into the demonstration project needs a powerful management with a strategic vision and a strong commitment to maximize the project benefits and to ensure the project continuation. This applies for innovative technology projects in general. Fuel cell buses are not an exception.

5

Hydrogen-Based Energy Market in Poland

M. Stygar and T. Brylewski

Faculty of Materials Science and Ceramics,
Department of Physical Chemistry and Modeling,
AGH University of Science and Technology, Kraków, Poland

5.1 Introduction

A rational energy policy and security and the diversification of energy sources are one of the significant issues in Poland. In recent years, the central authorities and a number of national institutions have made significant efforts to comply with the EU requirements concerning the energy policy.

At the same time, the notion of using hydrogen as an energy carrier remains marginalized in the context of the future energy industry in Poland. This is demonstrated by the current state of the Polish energy industry and its strong reliance on bituminous and brown coal. Despite these obvious observations, Poland is a country with a relatively substantial capacity to make the transition from energy production based on fossil fuels to an energy industry involving numerous and diverse energy carriers and, ultimately, to an energy industry based on hydrogen and renewable energy sources [1].

The most significant factors, which may facilitate this desirable transition, are described in detail in the following papers: Staśko D., Kaliski M.; Stygar M., Brylewski T.; Murray M. et al. [1–3]. The most important of these considerations include:

- Relatively small deposits of natural gas and petroleum, which are mostly imported. The high efficiency of fuel cell technology would allow the existing deposits to be utilized to a much larger extent by and to obtain more energy from the same amount of resources.

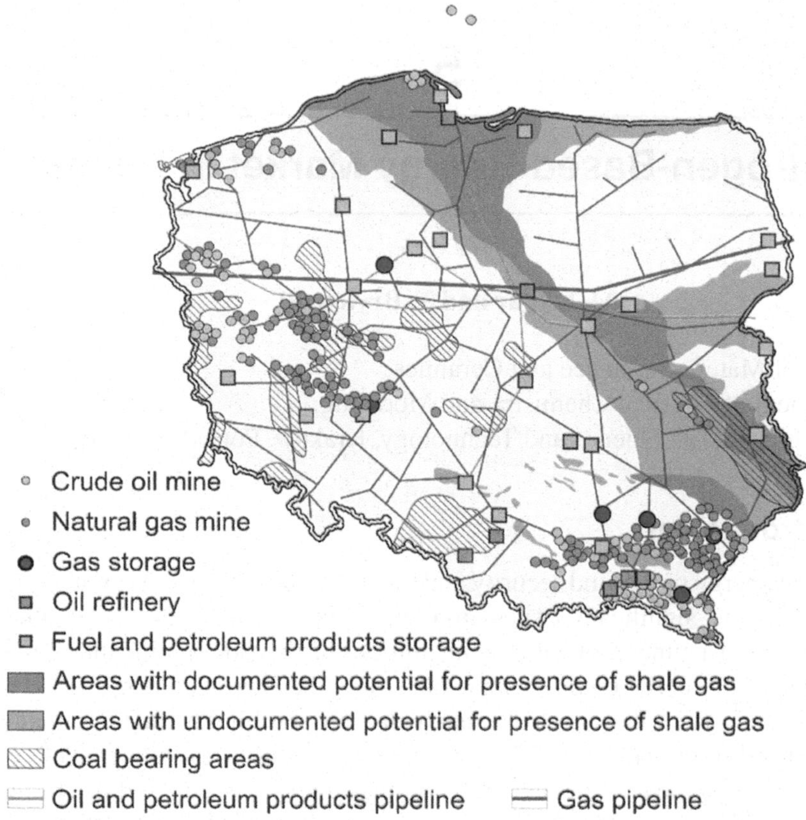

Figure 5.1 Polish natural resources and infrastructure with potential significance for hydrogen technology [4–10].

- Relatively uniform distribution of resources across the country, which would allow the production of fuel for the fuel cells on the site, eliminating the need to develop new power transmission infrastructure.
- a well-developed gas and electrical energy transmission infrastructure, numerous refineries, liquefied natural gas import terminals, underground gas storage facilities, and structures that allow the inexpensive production of hydrogen (e.g., chemical processing plants).

Moreover, the geological factors specific to the country suggest that one of the technologies used to produce hydrogen fuel should be the gasification of coal, which could provide enough energy to ensure Poland's energy self-sufficiency for several dozen years [11, 12]. The low calorific value of this fossil fuel, its contamination (predominantly with sulfur), and strong foreign competition might also affect a change in the strategy of utilizing this resource. For example, the much more efficient high-temperature solid oxide fuel cells (SOFCs) intended for application in on-site generation make it possible to utilize fuels based on natural sources such as natural gas, syngas, coal gas, biogas, or gas obtained via the gasification of light petroleum fractions [13]. Owing to the use of purification and enrichment systems—the so-called reformers—these systems can successfully be used with gases with varying levels of contamination [1, 14].

5.2 Energy Laws, Strategy and Policy of Poland

The "Energy Policy of Poland until 2030" announced in 2009 [15] set the current and future course for the Polish energy policy. It is worth noting that the goals set in this document are also compliant with the three main objectives of the policy of the IEA (International Energy Agency—an international organization that exerts influence over the global energy and fuel markets; Poland is a member of this organization since 2008), namely energy security, environmental protection, and economic development [16].

The aforementioned "Energy Policy of Poland until 2030"—a strategy formulated by the Ministry of Economy and approved by the government in November 2009—places the main focus of the energy policy on improving the energy security of the country while simultaneously ensuring sustained development [15].

The Polish energy policy entails the following main objectives:

- Improved energy security of the country;
- Increased security of fuel and energy supply;
- Diversification of energy generation via the adoption of nuclear power;
- Increased utilization of renewable energy sources, including biofuels;
- Development of competitive fuel and energy markets;
- Reducing the impact of power industry on the environment.

In addition, as part of Poland's obligations as an EU member state, the following goals were proposed to be achieved before 2020 [15, 17]:

- Reducing the emission of greenhouse gases in sectors that do not fall under the EU Emissions Trading System (EU ETS) to a level 14% higher than that recorded in 2005.
- Reducing the energy consumption levels projected for 2020 by 20%.

- Increasing the contribution of renewable energy sources to 15% of the final gross energy consumption, which includes increasing the contribution of renewable energy in transport to 10%.

Regardless of numerous declarations (and subsequent real and positive actions) concerning plans to increase the role of renewable energy sources in energy generation, the fact remains that in the documents approved by the Polish authorities, most notably the Energy Law of 10 April 1997 ("Prawo energetyczne") [18] and the Act on Renewable Energy Sources of 20 February 2015 ("Ustawa o odnawialnych źródłach energii") [19] completely omit the question of gradual transition from conventional power generation based on fossil fuels to a low-emissions energy production using devices such as fuel cells. It is worth keeping in mind that the definition of renewable energy is complementary with the notion of low-emissions energy production, but the latter one also encompasses fuel cells and other technologies, which are not necessarily based on renewable energy sources.

Hydrogen-based energy generation can be subsumed under the term "technological diversification" included in the "Energy Policy of Poland until 2030" [15] and considered a tool that can be used to achieve increased energy security. The document emphasizes the need to implement modern technologies in the energy sector and to intensify efforts connected with innovation in most fields, including the utilization of natural gas, district heating, and energy efficiency. It also formulates the requirement to "support the research and development of new technologies and solutions" in the field of energy efficiency, renewable energy sources, and clean coal technologies, including carbon capture and storage (CCS) mechanisms.

Although there are certain common points between the Polish research and development policy and the general energy policy (especially with regard to climate change and goals related to energy security), Poland has yet to develop a consistent research and development strategy with clearly defined objectives, long-term funding and monitoring, and the proper evaluation of the results.

5.3 Development of Hydrogen-Based Energy Generation—National Institutions

A number of national institutions have the competences and authority to promote, implement, and control hydrogen-based energy generation. However, the lack of a consistent policy makes it impossible to establish a single institution that would take full responsibility with regard to this matter and take over the duties described in the following section.

• **Ministry of Energy**—The most important institution responsible for the energy policy and security. Its structure includes the Undersecretary of State responsible for energy-related issues and a number of departments that deal with specific aspects of the energy policy: the Department of Energy (electrical energy, energy efficiency), the Department of Renewable Energy (co-generation, renewable energy sources), the Department of Petroleum and Natural Gas, the Department of Mining (coal), the Department of Nuclear Energy, and others. As of today, none of the aforementioned departments focuses specifically on the development and utilization of the potential of hydrogen-based energy in Poland. The policies of both the current and the former governments concerning this issue have not created any opportunities to introduce complex strategies of building and developing hydrogen infrastructure or to increase the awareness of the subject on the part of the society [20].

• **Energy Regulatory Office**—A regulatory body of the Polish energy sector. It is subordinate to the Minister of Energy, but it is independent when it comes to decision-making. The president of the URE is nominated by the Prime Minister of Poland at the request of the Minister of Energy. In the context of the potential development of hydrogen-based energy generation, the most important duties of the URE include the following: granting concessions for the production of electrical energy and heat, the transmission and distribution of natural gas and electrical energy, and the sale of fuels and electrical energy. The URE also has some other duties, such as issuing certificates of origin for electrical energy generated from renewable sources and via co-generation, monitoring energy and fuel markets, and promoting competition [17].

• **Ministry of Treasury**—Represents the government as the owner of companies that belong in part or in their entirety to the state. By formulating and implementing privatization strategies, the ministry would be able to significantly affect the development strategies of state-owned energy companies. Thus far, none of the companies that produce and distribute electrical energy have any development plans pertaining to this technology (even though they are developing technologies based on renewable sources, such as wind power) [21].

• **Ministry of Environment**—Responsible for the environmental aspects of the energy sector, including the emission of CO_2 and other greenhouse gases as well as ecological fees. It also issues concessions for searching for resources and their extraction. Moreover, it is responsible for the accreditation of the Polish Environmental Technology Verification Body. The Institute of Technology and Life Sciences (ITP) is a Polish verification body accredited for ETV. The Institute's objective is to conduct research and development together with the operational implementation, dissemination, advisory, educational, training, promotion, inventive, and monitoring activities in fields ranging from agricultural science and technology, water management, supply and sanitation, materials engineering, and renewable energy. Within the ITP, a separate organizational unit (JWTS) that reports directly to the Director having the office in Poznań carries on the EU Environmental Technology Verification (ETV) pilot program. The role of this unit is to support manufacturers and suppliers in their efforts to commercialize new environmental technologies. This is the first verification body in Poland that operates according to the PN-EN-ISO/IEC 17020 standard and the ETV General Verification Protocol (GVP) [22].

- **Energy Market Agency**—Joint-stock company that provides information on the energy sector and the environment to the government, international organizations (including the International Energy Agency), scientific institutions, universities, and companies. The company is contracted by the Ministry of Economy, which is responsible for data related to energy in Poland. It also prepares forecast for the energy sector at the request of the government [23].
- **Environmental funds**—Poland has a system of environmental funds with a structure that includes four levels. The National Fund for Environmental Protection and Water Management (Narodowy Fundusz Ochrony Środowiska i Gospodarki Wodnej, NFOŚiGW) finances activities related to environmental protection under the supervision of the Ministry of Environment; investments aimed at improving energy efficiency and the utilization of renewable sources have recently been prioritized. There are also 16 Regional Funds for Environmental Protection and Water Management, which operate as independent legal entities as well as environmental funds in counties and municipalities, which are governed by local authorities [24].

5.4 Development of Hydrogen-Based Energy Generation—Non-Governmental Organizations

Poland has a significant yet scattered research potential in the field of hydrogen-related technology, especially high- and low-temperature fuel cells as well as methods of hydrogen generation and storage. The most active organization of Polish specialists in this field is the Polish Hydrogen and Fuel Cell Association.

Moreover, two of several dozen clusters of companies, research organizations, business-related organizations, and public entities operating in Poland have declared the will to cooperate not only in the field of renewable energy but also in hydrogen-based energy in their statutes and corporate charters.

5.4.1 Polish Hydrogen and Fuel Cell Association

The Polish Hydrogen and Fuel Cell Association (PHAFCA), which had been established in June 2004, is the most significant organization that promotes hydrogen-based energy in the country. It comprises 176 regular members (including 42 professors), two honorary members, and 11 contributing members [25]. The organization is supported mainly from the contributions made by the latter and subsidies from the Ministry of Science and Higher Education. The association collaborates with many domestic and foreign institutions. It is a member of two prestigious international organizations that promote the development of hydrogen-based energy—the European Hydrogen Association (EHA) and The Partnership for Advancing the Transition to Hydrogen (PATH) – among others.

The fundamental activities of the association include creating a framework for the cooperation between scientists and the industry, promoting related technologies, and educating the society. It has organized the Polish "SMART ENERGY Conversion & Storage" Forum, the PHAFCA summer schools, a series of presentations and lectures aimed at reducing emissions in Kraków and promoting the use of electric vehicles in public transport ("Kraków—Stop Emisji: Elektryczna Komunikacja Miejska"), and it has also published the PHAFCA Bulletin.

In addition, the association promotes research on fuel cells and hydrogen-based technologies by staging an annual contest for the best MSc and PhD dissertations in the field.

5.4.2 Polish Hydrogen and Fuel Cell Technology Platform

The Polish Hydrogen and Fuel Cell Technology Platform was established on the 21st of January, 2005 [25, 26]. This organization brings together 40 entities from various sectors as well as research institutes and technical universities. The coordinator of the Platform is currently Prof. Jacek Kijeński from the Industrial Chemistry Research Institute.

Its most significant industry partners include companies such as PKN ORLEN S.A., Polskie Sieci Elektroenergetyczne S.A., Grupa Azoty S.A., Zakłady Azotowe Puławy S.A., Kompania Węglowa S.A., Południowy Koncern Energetyczny S.A., Polimex-Mostostal SIEDLCE S.A., and Consolidated Seven Rocks Mining Ltd. There are also 15 research units among its members, including the Warsaw University of Technology, the Silesian University of Technology, the Central Mining Institute, the Oil and Gas Institute, the University of Science and Technology (AGH-UST), three centers of advanced technologies, one center of excellence, the Polish Chamber of Chemical Industry, and the Polish Hydrogen and Fuel Cell Association.

Some of the objectives set by the members are improving the competitiveness of the Polish economy with regard to the production and utilization of hydrogen in the power industry, the chemical industry, the gas industry, and other sectors, with the use of fuel cells. They have also expressed the will to cooperate with the partners of the platform in fields connected with hydrogen technologies and to formulate a vision and strategy for developments related to the production and utilization of hydrogen in the power industry, the chemical industry, and agriculture.

5.4.3 Małopolska and Podkarpacie Clean Energy Cluster

The Cluster is an organization founded by the University of Science and Technology in Kraków (AGH-UST). It brings together universities, research units, state and private companies, and Marshall's Offices. Its main aim is the formation of a platform that would allow knowledge and information exchange between scientific and industrial communities as well as local government organizations. Several dozen seminars, meetings, and workshops have so far been organized as part of the Clean Energy Cluster's activities. Its strategic goals include the following:

- Allowing scientific institutions, universities, and companies to confront their expectations pertaining to new production technologies, energy conservation, and the reduction of adverse impact on the environment with the newest results and designs of innovative technologies.
- Promoting cooperation between entities.
- Obtaining knowledge on innovative hydrogen technologies and renewable energy sources—Applying innovative technologies and hybrid systems for power generation using renewable sources and the utilization of distributed energy sources.

The Cluster's activities are coordinated by the Department of Economics and Management in Industry, a unit of the Faculty of Mining and Geoengineering at the AGH-UST [25, 27, 28].

5.4.4 Baltic Eco-Energy Cluster

The Baltic Eco-Energy Cluster is a joint initiative of the Institute of Fluid-Flow Machinery of the Polish Academy of Sciences, the University of Warmia and Mazury in Olsztyn, the Gdańsk University of Technology, the Koszalin University of Technology, representatives of the local government, and economic entities and associations based in the corresponding voivodeships.

The cluster actively implements ideas of distributed co-generation in its broadest sense, that is, the simultaneous generation of thermal and electrical energy on a small and medium scale using renewable energy sources, especially biomass, biogas as well as energy harnessed from wind, the sun, water, hydrothermal energy, and hydrogen technologies [29].

5.5 Development of Hydrogen-Based Energy Generation—Research Units

In recent years, the organizations listed in Section 5.3 have contributed to the far-reaching consolidation of research communities and entities that conduct research and development activities in the field of sustainable energy, including clean coal technologies, renewable energy, materials technology, transport and environmental protection.

Nevertheless, in addition to activities conducted by research consortia, associations, or clusters, a number of universities and scientific institutes have been establishing separate, independent organizational units that engage in interdisciplinary research and development, service and training operations in the field of power generation in the broadest sense.

Such centers include:

- The Center of Energy at the University of Science and Technology in Kraków (AGH-UST) [30];
- The Academic Research Center for Power Engineering and Environment Protection at the Warsaw University of Technology [31];
- The Center of New Technologies (CENT) at the University of Warsaw [32].

Figure 5.2 Research in the field of hydrogen energy—research units and clusters.

5.6 Hydrogen-Based Energy in Poland—Conferences and Events

5.6.1 The Polish "SMART ENERGY Conversion & Storage" Forum

The Polish "SMART ENERGY Conversion & Storage" Forum is organized once in every two years by the Polish Hydrogen and Fuel Cell Association. By 2015, five such conferences had been organized.

The subject matter of the Forum includes issues associated with the production and storage of hydrogen, fuel cells, photoelectrochemical cells, photovoltaics, lithium cells, supercapacitors, and thermoelectric materials.

Table 5.1 Units that conduct research in the field of hydrogen energy generation (the number in the first column corresponds to the numbers in Figure 5.2)

No.	Unit Name	Location	General Research Areas
1	Warsaw University of Technology	Warsaw	Intelligent energy systems; Fuel cells; Modeling of energy devices; Fundamental research; Applied research
2	University of Warsaw	Warsaw	Hydrogen storage; Fuel cells; Lithium batteries
3	Institute of Power Engineering–Research Institute	Warsaw	Fuel cells; Coal conversion (gasification); Fundamental and applied research
4	Institute of Physical Chemistry PAS (Polish Academy of Sciences)	Warsaw	Fundamental research
5	Electrotechnical Institute	Warsaw	Fundamental research; Automatic control of energy systems; Fuel cells
6	Industrial Chemistry Research Institute	Warsaw	Electrochemical sources of energy; Fuel cells; Fundamental research
7	University of Science and Technology (AGH-UST)	Kraków	Intelligent energy systems; Fuel cells; Energy device modeling; Hydrogen storage; Lithium batteries; Fundamental research; Applied research
8	Oil and Gas Institute	Kraków	Gaseous fuels for application in the gas industry
9	Institute of Nuclear Physics PAS	Kraków	Fundamental research
10	Institute of Molecular Physics PAS	Poznań	Fundamental research
11	Institute of Non-ferrous Metals Division in Poznań–Central Laboratory of Batteries and Cells	Poznań	Fuel cells; Hydrogen storage; Lithium batteries; Hybrid systems
12	Poznan University of Technology	Poznań	Distributed generation; Fuel cells
13	Adam Mickiewicz University in Poznań	Poznań	Fundamental research; Biogeneration
14	Institute of Low Temperature and Structure Research PAS	Wroław	Fundamental research; Fuel cells
15	Gdańsk University of Technology	Gdańsk	Fuel cells

(*Continued*)

Table 5.1 Continued

No.	Unit Name	Location	General Research Areas
16	Częstochowa University of Technology	Częstochowa	Production of fuel for fuel cells; Fuel cells; Applied research
17	Institute for Chemical Processing of Coal	Zabrze	Production of hydrogen; Production of biofuels
18	Institute of Chemical Engineering PAS	Gliwice	Production of hydrogen; Reforming methanol for fuel cells
19	West Pomeranian University of Technology	Szczecin	Smart grids; Production of hydrogen; Fundamental research
20	Silesian University of Technology	Katowice	Fuel cells; Energy system modeling
21	Institute of Power Engineering–Ceramic Department CEREL	Rzeszńw	Fuel cells
22	Lublin University of Technology	Lublin	Fuel cells

The conferences presents the newest achievements in research conducted in Poland and the most renowned foreign scientific institutions [25].

5.6.2 EU Sustainable Energy Week

The EU Sustainable Energy Week held since 2013 at the University of Science and Technology in Kraków (AGH-UST), among others, is an initiative that is a part of the cycle of such events organized annually across Europe in the second half of June [25, 33, 34]. The European Commission is the initiator of this campaign. Each year, after all associated events have ended, the Sustainable Energy Europe Awards are given to the best projects that promote the reduction of energy consumption, a sustainable energy policy, and so on.

The goal of the Sustainable Energy Week is to promote the idea of sustainable development and its energy-related, ecological and economical aspects, the exchange of experience and knowledge, and to strive for the development and widespread adoption of new technologies associated with energy and ecology. In addition, one of the aims of the Sustainable Energy Week is making the society more aware of energy- and ecology-related concerns and proposing good practices and state-of-the-art solutions.

The principal items on the Sustainable Energy Week agenda include: a science conference, the exhibition of devices manufactured by leading companies in the renewable energy sector as well as HVAC and the related sectors, the presentation of the so-called energy park that involves experiments and scientific shows, workshops and expert training, open days for schools, and an open debate. With each edition, the agenda of the Sustainable Energy Week is supplemented with new, interesting events.

5.6.3 International Scientific Conference on Sustainable Energy and Environment Development SEED'2016

The SEED conference, a key event of the 2016 Sustainable Energy Week, was first held between 17 and 19 May 2016 [34, 35]. This event replaced the First Polish Scientific Conference on Energy and Fuels (I Ogólnopolska Konferencja Naukowa "Energia i Paliwa") organized as part of the 2015 Sustainable Energy Week. The subject matter encompasses issues connected with the power industry, the production and storage of energy, conventional and alternative fuels, energy-efficient buildings, smart city and smart grid concepts, and environmental protection in the power industry and waste management.

5.6.4 Energy–Ecology–Ethics Student Conference

The Polish Energy–Ecology–Ethics Scientific Conference (Ogólnopolska Konferencja Naukowa "Energia–Ekologia–Etyka") has been organized since 2002 by the students of the University of Science and Technology in Kraków (AGH-UST) [34, 36]. In 2013, it was officially incorporated into the Sustainable Energy Week.

The Conference involves a wide scope of issues connected with energy generation, environmental engineering and protection, and ethics in science and business. It is addressed to students and PhD students from the entire country, and it provides an opportunity to present one's scientific achievements.

5.6.5 Kraków—Stop Emisji: Elektryczna Komunikacja Miejska

The unprecedented "Kraków—Stop Emisji: Elektryczna Komunikacja Miejska" event, which has so far been organized just once, was held on 23 May 2014 at the University of Science and Technology in Kraków (AGH-UST). Its goal was to present environmentally friendly solutions as an alternative to cars to the inhabitants of Kraków. The event was organized by the Polish Hydrogen and Fuel Cell Association, the University of Science and Technology in Kraków (AGH-UST), and the Hydrogenium research circle at the Department of Hydrogen Energy that operates as part of the AGH-UST Faculty of Energy and Fuels.

During the event, experts delivered lectures on the future developments in the Li-ion battery technology and the safety of their application in the automotive industry, and achievements in the field of supercapacitors and thermoelectric materials—technologies important for the production of electric and hybrid vehicles. The event also featured speeches by representatives of the city authorities and the University, meetings with representatives of companies that manufacture batteries for the global market, a demonstration of electric and hybrid cars produced by Mitsubishi, Nissan, and Toyota, and the first electric bus owned by Kraków's public transport operator [25].

5.7 Scientific Initiatives in the Field of Hydrogen Energy—Research Grants, Projects, and Investigations

The research and development work performed thus far in the field of hydrogen-based energy generation in Poland can be considered insufficient. A number of polish scientific institutions, universities, and companies have the resources and the scientific and research know-how required to develop hydrogen technologies and to introduce them to the market. In the span of the last dozen or so years, several dozen domestic grants and a dozen or so large grants as part of the international cooperation were obtained from various sources [37]. Unfortunately, due to the lack of a comprehensive national program for the development of hydrogen technologies, ready-made and likely much more expensive solutions may have to be purchased in other countries in the future [11].

The domestic sources of funding scientific research are currently the following:

- **The National Science Center**—A unit that supports scientific activity related to fundamental research, namely experimental or theoretical work undertaken primarily to gain new insight into the background of observed phenomena, without focusing on direct commercial application [38].
- **The National Center for Research and Development**—The Center's activity is geared toward enhancing cooperation between Polish businesses and to allow widespread commercialization of the results of scientific research in order to bring profit to the Polish economy.

Table 5.2 The most significant conferences and events

Name	Event Date	Frequency	Organizing Institution
Polish "SMART ENERGY Conversion & Storage" Forum	September	every two years (most recently in 2015)	- PHAFCA - AGH-UST—Faculty of Energy and Fuels - IATI (Institute of Highway Technology and Innovation)
EU Sustainable Energy Week	June	annually (since 2013)	- AGH-UST—Faculty of Energy and Fuels - AGH-UST—Center of Energy
International Scientific Conference on Sustainable Energy and Environment Development SEED'2016	17–19 May 2016	first event	- PHAFCA - AGH-UST—Faculty of Energy and Fuels
"Kraków—Stop Emisji: Elektryczna Komunikacja Miejska"	23 May 2014	one-time event	- AGH-UST—Faculty of Energy and Fuels - Foundation Institute for Sustainable Energy
PHAFCA summer/autumn schools	September or October	every two years (most recently in 2013)	- PHAFCA
Energy–Ecology–Ethics Student Conference	June	annually (since 2002, as part of the Sustainable Energy Week starting from 2013)	- AGH-UST—Faculty of Energy and Fuels - STN Students' Research Association affiliated with AGH-UST - "Eko-Energia" research circle

These goals are implemented via programs that support applied research and R&D, by funding commercialization efforts and the transfer of results to the economy, and by supporting the development of young scientists [39].

- **The Ministry of Science and Higher Education**—Governs the public administration departments and handles the budget for state-funded scientific research. In addition to allotting the budgets to the aforementioned centers, it organizes and independently handles a certain number of research grants [40].

After the accession of Poland into the EU, there emerged new opportunities to gain funding as part of the EU's Horizon 2020 program as well as various operational programs and other projects.

The Foundation for Polish Science [41], which has been active since 1991, is also worthy of note. It is a non-governmental, non-profit organization that supports scientific development. It is the largest non-state source of research funding in Poland.

Thus far, the most significant results with regard to the development of comprehensive R&D programs have been achieved by the National Center for Research and Development. The year 2008 marked the first time when the Center designated the "Energy and Infrastructure" area as a strategic objective of R&D activities. In the past, this issue was scattered across many different projects and programs and had little strategic significance.

The "Energy and Infrastructure" research area currently encompasses the following research priorities:

- Reducing energy consumption through research and by implementing energy-saving solutions in the industry, the services sector, and in households.
- An effective and environmentally friendly utilization of the domestic fossil fuel resources.
- The development of alternative sources of energy and their carrier (renewable sources, nuclear energy, and hydrogen technologies).

The active participation of Polish scientists in EU structures as well as European consortia consisting of scientific and economic entities presents opportunities of obtaining funds for educating new leaders in science and the economy, and to achieve noteworthy results in interdisciplinary scientific research connected with renewable and low-emissions energy, among others. Past and present examples of such programs include:

- **Interdisciplinary doctoral studies—Advanced materials in modern technologies and the future power industry**

In the period from 2009 to 2015, the Faculty of Physics and Applied Computer Science at the University of Science and Technology (AGH-UST)—the leader of the project, together with its partners—the Institute of Nuclear Physics and the Institute of Catalysis and Surface Chemistry (both units of the Polish Academy of Sciences in Kraków), organized doctoral studies entitled Interdisciplinary Doctoral Studies—Advanced Materials in Modern Technologies and the Future Power Industry [42]. These studies were

Table 5.3 Examples of research grants from various sources of funding

Grant Title	Type of Grant	Beneficiary
Source of funding—National Science Center		
Functional layers of solid oxide fuel cells.	OPUS—Funds for purchasing or designing research equipment.	Gdańsk University of Technology
Metal/ceramics layered systems designed for application in SOFCs in Ar-H_2-H_2O/air dual reaction atmosphere.	PRELUDIUM—Funding intended for pre-doctoral researchers about to embark on their scientific career.	University of Science and Technology (AGH-UST)
Novel electrode materials for IT-SOFCs fueled by syngas.	ETIUDA—Funding intended for Ph.D. candidates.	University of Science and Technology (AGH-UST)
Source of funding—Ministry of Science and Higher Education		
Increasing the corrosion resistance of porous metallic materials applied in solid oxide fuel cells.	IUVENTUS PLUS	Gdańsk University of Technology
Source of funding—The National Center for Research and Development		
EkoRDF–An innovative method of obtaining alternative fuel from municipal waste.	GEKON Programme—Generator of Ecological Concepts.	A consortium of the University of Agriculture in Krakow, the Institute for Chemical Processing of Coal, and others.
Source of funding—Framework Programme: FP7 (EU grants)		
"Innovative Dual mEmbrAne fueL-Cell" (IDEAL-Cell).	FP7-ENERGY-2007-1-RTD	A consortium of the University of Science and Technology (AGH-UST) and others.
Development of PEM fuel cell stack reference test procedures for the industry.	FCH-JU-2011-1	A consortium of the Industrial Chemistry Research Institute and others.

co-financed by the European Commission and the Ministry of Science and Higher Education as part of the Human Capital Operational Programme, priority IV focused on higher education and science, Section 5.1. The intermediary institution was the National Center for Research and Development. Education and scientific research performed as part of the studies concerned the physical, chemical, and technological aspects of knowledge on materials and the modern power industry.

- **KIC InnoEnergy project**

This project was qualified for funding in 2009 by the European Institute of Innovation and Technology (EIT) in Budapest [43]. It was implemented by an international consortium coordinated by the Karlsruhe Institute of Technology (KIT)—one of the eight flagship German universities. The consortium consists of six nodes called Colocation Center (CCs), which are responsible for individual subject matter areas. The entire consortium forms the so-called Knowledge and Innovation Community (KIC). The coordinator of the Polish node (CC PolandPlus) is the University of Science and Technology (AGH-UST).

One of the aspects of this enterprizes is an array of doctoral studies [44], internships, and opportunities to conduct research at the institutions that belong to the consortium. One of the objectives of the project is education with a focus on practical application and the promotion of entrepreneurship among scientists.

5.8 Review of the Most Important Papers and Journals

Currently, there are no Polish journals that are dedicated to hydrogen-based energy and included on the ISI Master Journal List, or published in foreign languages. A similar situation is observed in journals dedicated to particular disciplines; however, despite the lack of a journal specifically devoted to hydrogen-based technologies, articles on hydrogen-based technology can be found in journals such as Energetyk, Inżnieria i Budownictwo, GLOBEnergial AgroEnergetyka, or Czysta Energia.

Each year, the Polish Hydrogen and Fuel Cell Association publishes the PHAFCA Bulletin [25]. It features topics connected to fuel cells and hydrogen-based technologies as well as other related issues in the field of energy generation. The Bulletin is on the list of journals scored by the Ministry of Science and Higher Education, with one point. Its editor-in-chief and subject editor is Professor Janina Molenda. It is currently the only Polish periodical dedicated exclusively to research on hydrogen-based energy generation.

Acknowledgment

The authors would like to kindly thank Dr. Andrzej Kacprzak—the founder of the Internet site *ogniwa-paliwowe.info*—for the opportunity to make use of the materials he had collected.

References

[1] Staśko D., Kaliski M. An evaluation model of energy safety in Poland in view of energy forecasts for 2005–2020. Arch. Min. Sci. 2006; 51: 311–46.

[2] Stygar M., Brylewski T. Towards a hydrogen economy in Poland. Int. J. Hydrogen Energy. 2013; 38: 1–9.

[3] Murray M., Seymour E., Rogut J., Zechowska S., Stakeholder perceptions towards the transition to a hydrogen economy in Poland. Int. J. Hydrogen Energy. 2008; 33: 20–7.

[4] Paska J., Ekonomika w elektroenergetyce [in Polish]., Warsaw: Oficyna Wydawnicza Politechniki Warszawskiej; 2007.

[5] International Energy Agency e IEA. Accessed: 1 March 2016. Available online at: (http://www.iea.org/).

[6] European Commission—Eurostat. Accessed: 1 March 2016. Available online at: (http://epp.eurostat.ec.europa.eu).

[7] Kulagowski W. Hydropower engineering in Poland—present state and development perspectives. Water Manage. 2001; 3.

[8] Steller K., Steller J. Research and development activity on small hydropower in Poland. Energ. Sourc. 1993; 15: 37–49.

[9] Institute of Meteorology and Water Management—IMGW. Accessed: 1 March 2016. Available online at: (http://www.imgw.pl/).

[10] The European Commission's In-house Science Service, Joint Research Center—JRC. Accessed: 1 March 2016. Available online at: (http://ec.europa.eu/dgs/jrc/).

[11] Molenda J. The fundamental meaning of scientific research for development of hydrogen economy. Energy Policy Journal. 2008; 11: 61–68.

[12] Energy Market Information Center (CIRE). Accessed: 1 March 2016. Available online at: (http://www.cire.pl/).

[13] Tomczyk P. Prospects and obstacles of hydrogen economy development. Energy Policy Journal 2009;12: 593–608.

[14] Singhal S. C., Kendall K. High Temperature Solid Oxide Fuel Cells. Fundamentals, design and applications. Oxford: Elsevier; 2003.

[15] Energy Policy of Poland until 2030. Accessed: 1 March 2016. Available online at: (http://www.mg.gov.pl/files/upload/8134/Polityka%20energetyczna%20ost.pdf).

[16] International Energy Agency e IEA. Accessed: 1 March 2016. Available online at: (http://www.iea.org/).

[17] Energy Regulatory Office. Accessed: 1 March 2016. Available online at: (http://www.ure.gov.pl/).

[18] Polish Parliament—ISAP: Internet System of Legal Acts, The Polish Energy Law. Accessed: 1 March 2016. Available online at: (http://isap.sejm.gov.pl/DetailsServlet?id=WDU19970540348).
[19] Polish Parliament—ISAP: Internet System of Legal Acts, Act on Renewable Energy Sources. Accessed: 1 March 2016. Available online at: (http://isap.sejm.gov.pl/DetailsServlet?id=WDU20150000478).
[20] Ministry of Energy. Accessed: 1 March 2016. Available online at: (http://www.me.gov.pl/).
[21] Ministry of Treasury Republic of Poland. Accessed: 1 March 2016. Available online at: (http://www.msp.gov.pl/).
[22] Ministry of the Environment. Accessed: 1 March 2016. Available online at: (http://www.mos.gov.pl).
[23] Energy Regulatory Office. Accessed: 1 March 2016. Available online at: (http://www.are.waw.pl/).
[24] National Fund for Environmental Protection and Water Management. Accessed: 1 March 2016. Available online at: (http://www.nfosigw.gov.pl/).
[25] Polish Hydrogen and Fuel Cells Association. Accessed: 1 March 2016. Available online at: (http://www.hydrogen.edu.pl/).
[26] Polish Hydrogen and Fuel Cell Technology Platform. Accessed: 1 March 2016. Available online at: (http://www.ichp.pl/polska-platforma-technologiczna-wodoru-i-ogniw-paliwowych).
[27] Malopolsko-Podkarpacki Clean Energy Cluster. Accessed: 1 March 2016. Available online at: (http://klaster.agh.edu.pl/strona-glowna/).
[28] Podkarpackie Voivodeship Office. Accessed: 1 March 2016. Available online at: (https://rzeszow.uw.gov.pl/en/).
[29] Baltic Eco-Energy Cluster. Accessed: 1 March 2016. Available online at: (http://www.imp.gda.pl/bkee/).
[30] Center of Energy AGH. Accessed: 1 March 2016. Available online at: (http://www.agh.edu.pl/centrum-energetyki/).
[31] Uczelniane Centrum Badawcze Energetyki i Ochrony środowiska (Academic Center for Energy and Environmental Policy Research). Accessed: 1 March 2016. Available online at: (www.eios.pw.edu.pl/).
[32] Center of New Technologies University of Warsaw. Accessed: 1 March 2016. Available online at: (http://www.cent.uw.edu.pl/).
[33] EU Sustainable Energy Week. Accessed: 1 March 2016. Available online at: (http://eusew.eu/).
[34] Sustainable Energy Week at AGH University of Science and Technology. Accessed: 1 March 2016. Available online at: (http://www.tze.agh.edu.pl/).

[35] International Scientific Conference on Sustainable Energy and Environment Development SEED'2016. Accessed: 1 March 2016. Available online at: (http://www.seed.agh.edu.pl/).
[36] Student Conference—Energy-Ecology-Ethics. Accessed: 1 March 2016. Available online at: (http://www.eee.agh.edu.pl/).
[37] PAP Science & Scholarship in Poland. Accessed: 1 March 2016. Available online at: (www.naukawpolsce.pap.pl/).
[38] National Science Center–Poland. Accessed: 1 March 2016. Available online at: (http://www.ncn.gov.pl/).
[39] The National Center for Research and Development. Accessed: 1 March 2016. Available online at: (http://www.ncbir.pl/).
[40] Ministry of Science and Higher Education. Accessed: 1 March 2016. Available online at: (http://www.nauka.gov.pl/).
[41] Foundation for Polish Science (FNP). Accessed: 1 March 2016. Available online at: (http://www.fnp.org.pl/).
[42] PhD studies program: "Interdyscyplinarne Studia Doktoranckie—Zaawansowane Materiały dla Nowoczesnych Technologii i Energetyki Przyszłości"—supported by Human Capital Operational Program POKL.04.01.01-00-434/08-02 co-financed by the EU. Accessed: 1 March 2016. Available online [in Polish] at: (http://www.isd.fis.agh.edu.pl/).
[43] KIC InnoEnergy. Accessed: 1 March 2016. Available online at: (www.kic-innoenergy.com/; http://www.smartgrid.agh.edu.pl/index.php/kic/).
[44] KIC InnoEnergy PhD School. Accessed: 1 March 2016. Available online at: (http://www.kic-innoenergy.com/phdschool/).

6

The Hydrogen-Based Energy Aspects in Romania

Ioan Ştefãnescu[1,2] and Ioan Iordache[1,2]

[1]National Research and Development Institute for Cryogenics and Isotopic Technologies ICSI, Rm. Valcea, Romania
[2]Romanian Association for Hydrogen Energy, Rm. Valcea, Romania

6.1 Introduction

Romanian's energy system future is in debate and the European context will put its mark. The hydrogen economy is one of the future energy system components. The European and national strategies on the development of hydrogen and fuel cell technologies vary considerably from country to country. Two types of policies were proposed: policies of European Commission, supported by Fuel Cells and Hydrogen Joint Undertaking (FCH JU), and policies of Members States, both national and regional. In the past years, Romania has made considerable progress in developing institutions compatible with a market economy and joining the European Union (EU) initiatives for a hydrogen economy. In contrast to USA (e.g., 15 national laboratories, nearly 300 university research groups, and powerful companies) actively coordinated by DoE, the European activities on hydrogen and fuel cells are fragmented and not well coordinated [1].

The contribution of energy sources in Romania is a consequence of the past development of electric power industry, ex-socialist centrally planned economy. Romania's domestic energy production from coal, lignite, oil, gas, and hydropower covers around 70% of the energy needs [2]. The energy laws promote the production of energy from renewable sources, via "green certificates". Renewable energy sources are subsidized through the Romanian Environmental Fund and the National Rural Development Program [3]. The renewable energies include non-fossil energy sources: wind, solar, geothermal, hydrothermal and ocean energy, hydro, biomass, landfill gas, known as gas storage and gas from sludge digestion sewage treatment plants and biogas. The target ranged from 17.1% in 2006 to 23.4% in 2010, and the target is 24% by 2020. Romania is one of the countries with the biggest share of renewable energy, above the EU average. Romania was the second to increase the share of renewables. Estonia recorded the highest increase in its share from 16.1% to 24.3%, from 2006 to 2010 [4]. The fastest developing types of renewable resource are currently wind farms, the total power output produced, according to 2013 statistics, was 4721 GWh, as well as biomass power plants, which generate electricity from biomass. In 2013, their total power output was 314 GWh. Despite the relatively favorable conditions, Romania was generally not included in global statistics, since the total solar power output (i.e., electricity generated using photovoltaic cells as opposed to thermal energy obtained using solar thermal collectors) by the end of 2013 was equal to 413 GWh, according to the national transmission system operator, Transelectrica SA.

EU dependency on energy imports increased in the past years. Among the other Member States, lower dependency rates were recorded in Romania, 21.7% in 2010. The same can be said about the oil consumption. Dependency on oil import was only 51.3%, compared to the EU average 84.3%. This statement is true for natural gas dependence as well. EU dependency on natural gas grew by 13%, from 48.9% in 2000 to 62.4% in 2010. In 2010, the lowest natural gas dependency rate from UE was recorded in Romania (16.8%). Statistics show that at EU level Romania is an exporter of electricity. In the past decade, Romania thermal efficiency decreased slightly with the a few percentages, in contrast with the EU trend. Combined heat and power generation in Romania declined by nearly half in comparison with 2005 when was recorded 26.2% of gross electricity generation. On the contrary, in the EU made up 11.7% of gross electricity generation. Energy intensity in Romania recorded a decrease of 35 % between 2000 and 2010, much more than the EU average, 10.3%. The electricity consumption *per capita* in Romania grew continuously. At sector level, the largest growth was observed in services and transport, whereas energy consumption by the residential sector and industrial sector declined.

Table 6.1 shows the indicators that are possible to have an influence in terms of transition to the hydrogen economy. Renewables will have an important share in the energy sector in the context of hydrogen economy. In Romania, hydro covers an important part in the production of electricity; other renewables are in a relatively increased development. CO_2 emissions are relatively constant; decreasing and increase periods are alternating [5].

Table 6.1 Romanian's socioeconomic development indicators

Indicator	2000	2001	2002	2003	2004	2005	2006	2007	2008	2009	2010
Alternative and nuclear energy (% of total energy use)	7.46	7.36	7.43	6.18	7.45	8.29	7.68	8.57	11.20	12.71	13.60
CO_2 emissions (kt 10^{-3})	89.98	95.67	92.17	95.94	95.40	94.96	102.0	99.70	94.77	79.49	
CO_2 emissions from electricity and heat production, total (% of total fuel combustion)	56.74	55.79	54.70	55.70	52.60	51.43	51.99	51.84	51.60	51.35	51.01
Electricity production (TWh)	51.93	53.87	54.74	55.14	56.50	59.41	62.70	61.67	64.96	57.74	60.26
Electricity production from oil, gas, and coal sources (% of total)	61.04	62.19	60.61	67.05	60.95	56.63	61.73	61.55	56.20	52.71	47.62
Electricity production from hydroelectric sources (% of total)	28.46	27.70	29.32	24.05	29.23	34.01	29.28	25.89	26.47	26.90	32.40
Electricity production from renewable sources, excluding hydroelectric (% of total)	0.00	0.00	0.01	0.01	0.01	0.01	0.01	0.06	0.04	0.03	0.69
Electricity production from nuclear sources (% of total)	10.51	10.11	10.07	8.90	9.82	9.35	8.98	12.50	17.28	20.35	19.29
Energy imports, net (% of energy use)	21.77	24.56	23.33	25.53	26.63	27.80	29.92	30.17	26.74	18.55	21.57

(Continued)

Table 6.1 Continued

Indicator	2000	2001	2002	2003	2004	2005	2006	2007	2008	2009	2010
Energy production (kt of oil equivalent 10^{-3})	28.31	27.76	29.23	29.36	28.36	27.91	27.95	27.72	28.98	28.34	27.44
Fossil fuel energy consumption (% of total)	84.81	86.53	86.74	86.84	84.48	83.90	85.04	83.15	79.81	76.54	75.13
Researchers in R&D (per million people)	923	893	922	956	973	1054	945	869	898	895	
Research and development expenditure (% of GDP)	0,37	0,39	0,38	0,39	0,39	0,41	0,45	0,53	0,59	0,48	
Road density (km of road per 100 sq. km of land area)						33,50	33,50	33,90	34,30	34,28	34,30
Road sector energy consumption (% of total energy consumption)	7,45	9,71	9,40	9,76	10,38	9,96	10,02	10,19	11,77	13,76	12,47
Road sector diesel fuel consumption (kt of oil equivalent)	1374	1815	1902	2152	2174	2130	2421	2400	3160	3166	3028
Road sector gasoline fuel consumption (kt of oil equivalent)	1232	1578	1540	1564	1623	1539	1438	1449	1448	1437	1286
Roads, total network (km 10^{-3})	198.6	198.6	198.9	198.7	198.8	198.8	198.8	198.8	198.8	198.8	198.8
Motor vehicles (per 1,000 people)				177	185	180	172	194	219	230	235
Technicians in R&D (per million people)	292	269	292	248	253	230	207	201	214	185	

Romania imports hydrocarbons; the trend is constant in the past decade. Regarding oil, Romania consumes ranging from 192 to 224 thousand barrels per day, but worse is that domestic production decreases constantly. The oil and natural gas proved reserves also declined in the recent decades, and Romanian coal available on the market is of poor quality [6].

The declared objective of Romania's energy policy is to ensure energy security, which guarantees continuous, stable economic growth and the ability to cover the full domestic consumption of electricity and heat under sustainable constraints (e.g., reduction in greenhouse gas emissions end decarbonization, reduction in particulate emissions, etc.). The requirements include also the gradual liberalization of electricity markets and a policy promoting sustainable development, for example, by more rational use of generated energy and the reduction in energy consumption. These aims are referred to as the 20-20-20 policy, which entails a 20% reduction in greenhouse gas emissions, a 20% reduction in energy consumption, and a 20% increase in the contribution of renewable sources, for the final requirement Romania wants to reach 24%. We underline that the Romanian's energy system has followed its own development strategy in line with the country's own needs and also influenced by European energy policy. In the past 20 years, Romania has made a considerable progress developing institutions compatible with a market economy and joining the European Union (EU) initiatives toward harmonized energy system transition, for example, hydrogen as an energy carrier.

In Romania, renewables will remain an important share for electricity generation/balancing, also in the context of an energy system applying hydrogen as energy vector. The hydropower covers not only a large share of the electricity production but also other renewable energy sources are increasingly being tapped. The fastest developing renewable energy technology is wind energy, extending the net electricity generating capacity. In the period of 2014–2015, its share was between 12.7%–15.3% of the Romanian net generating capacity. The gross renewable generating capacity and installed power in 2013 was hydro 15,104 GWh (6,648 MW), wind 4,721 GWh (2,607 MW), PV 413 GWh (860 MW), and biomass 319 GWh (96 MW) [7].

6.2 The Hydrogen Economy Perspectives in Romania

The Romanian authorities have not yet adopted any program dedicated to hydrogen and fuel cell that would specify objectives and assignments, integrate and coordinate the individual activities. Germany, Norway, Denmark, USA,

Japan, Canada, or South Korea are leaders in the implementation of hydrogen and fuel cell technology. They have already developed national directions for hydrogen implementation and have already in use production and distribution of hydrogen, hydrogen vehicles, and a network of the refueling stations. The leadership of these countries can be recognized by the implemented hydrogen technology and national strategy documents. Comparing countries on the basis of the amount of implemented hydrogen technology is very hard, because this technology is still either in the development or in the demonstration stage. Romania is a country with good opportunities to make the transition from the dependence on fossil fuels, to a power industry based on diverse energy carriers (such as hydrogen, among others), and to a power sector utilizing hydrogen and renewable energy sources. The decision to replace the actual systems would offer an opportunity to implement fuel cell stacks and to use hybrid systems that combine conventional and modern methods of electricity generation. The Romanian stakeholders would be eligible for subsidies from the EU funds. By adopting hydrogen-based technologies and thus reducing greenhouse gas emissions, the actors from Romania would be able to make profits or to recover their investments via the so-called emissions trading.

In Romania there are little more than 100 research institutes and universities with research laboratories. However, only a fraction of them are involved in the field of hydrogen and fuel cell research, development, and innovation. The hydrogen initiators have been conducting investigations on a number of issues related to the topic of hydrogen and fuel cell technologies. These works concern the generation and storage of hydrogen, fuel cell construction, and their electric and transport properties. Public funding on research and development is available through a national authority subordinated to the relevant ministry. A gross classification indicates that, in Romania, the hydrogen and fuel cell projects were included in research areas of energy and materials. The projects are few in number and they cover a significant broad research area, giving the impression of an unsynchronized activity. In a slight contrast with the European reality, the projects conducted in Romania are targeted more toward research and less on development or demonstration. The researchers are well inspired and focus on maximum use of the existing potential, for example, the use of H_2S in the Black Sea area; it should be remembered that the subject is actively addressed in several countries. Other researchers from Romania are turning on more defined themes such as membranes of palladium/ceramic, others are baked in multidisciplinary projects that relate to the production of hydrogen through the thermochemical Cu-Cl cycle, taking into account Romania's nuclear energy infrastructure [8].

A part of the main institutes or groups involved in the research, development, and innovation of hydrogen and fuel cells technologies are mentioned here. In the National Hydrogen and Fuel Cell Center, the teams of researchers are being involved in projects that involve: production, storage, and use of hydrogen in fuel cell hybrid systems, studying new materials for fuel cells and hydrogen technology, energy applications of hydrogen and fuel cell technology. In the National Institute for Research and Development of Isotopic and Molecular Technologies Cluj-Napoca, the researchers study materials for hydrogen storage, since the '80s. Similar activities for hydrogen storage in hydrides are taking place at the National Institute of Research and Development for Technical Physics from Iasi. The National Institute for Research and Development in Electrical Engineering, Bucharest, works on the integration of fuel cells in various electrical applications. New materials for fuel cells, particularly SOFC, can be identified at the National Institute of Materials Physics, Magurele. Ceramic materials and their potential use for hydrogen and fuel cell technology are tested at the Institute of Physical Chemistry of the Romanian Academy. The hydrogen production from biomass is developed at the Institute of Conception, Research and Design for Thermal Energy Equipment (OVM). The research about hydrogen and fuel cells technologies takes place at a series of universities also: *Politehnica* University of Bucharest (materials for solid oxide fuel cell, hydrogen for energy system and non-conventional sources of hydrogen), University of Bucharest (nanotechnology applications in hydrogen and fuel cell technology), Ovidius University of Constanta (Center for Advanced Engineering Sciences looking for the Black Sea area, and its potential for hydrogen economy), Transylvania University of Brasov (the production of hydrogen by water photolysis), Technical University of Cluj-Napoca, Polytechnic University of Iasi, Petroleum-Gas University of Ploiesti, University of Pitesti and Babes-Bolyai University.

In the past, institutes from Romania have participated in European Hydrogen and Fuel Cell Platform (HFP) and HY-CO ERA-NET project, when was established the Romanian Hydrogen and Fuel Cell Technology Platform, in 2004–2005. There were involved 11 entities: research institutes, universities, and SMEs. Leader of the consortium was the National Research Development Institute for Cryogenic and Isotopic Technologies—ICIT Rm. Valcea, where was created the National Center for Hydrogen and Fuel Cell in 2009.

Using Scopus, a bibliographic database containing abstracts and citations for academic journal articles, we are found over 100 works (articles and reviews) in the field of fuel cell and hydrogen energy from Romania covering subject areas such as: engineering, materials science, chemistry and chemical engineering, energy, environmental science, and mathematics, since 2000. As an overview, Romania is positioned somewhere in the middle of the ranking if we make a comparison with nearby Eastern European countries. Countries such as Russia, Turkey, and Poland have more than 300 scientific papers in this field, Greece and Austria with more than 100 scientific papers; Hungary, The Czech Republic, Ukraine, and Bulgaria are presented with a small proportion, less than 50 articles per country.

Two institutes from Romania, National Center for Hydrogen and Fuel Cells and OVM-ICCPET Institute (OVM), participated in the FCH JU financed projects: "Assessment of the potential, the actors and relevant business cases for large scale and seasonal storage of renewable electricity by hydrogen underground storage in Europe" and "Molten Carbonate Fuel Cell Catalyst and Stack Component Degradation and Lifetime: Fuel Gas Contaminant Effects and Extraction Strategies". Activities relate to a case study concerning the underground storage of hydrogen, (feasibility, relevance, timelines, chances, and limitations of H_2 underground storage to facilitate renewable electricity in Europe): and investigation of poisoning mechanisms caused by alternative fuels and applications and determining precisely MCFC tolerance limits for long-term endurance, and optimizing fuel and gas cleaning to achieve tailored degrees of purification according to MCFC operating conditions and tolerance.

The hydrogen is used mainly by the chemical industry, in refineries and for ammonia production, and production has so far been dominated by reforming of hydrocarbons, pyrolysis, and co-pyrolysis. But the hydrogen will become an energy vector along with electricity. The hydrogen is the most abundant element in the universe and can be obtained from a number of resources using various processes in the future more and more dominated by renewable technologies, and technologies must take into account both aspects of ecology and economy. In Romania were identified 13 industrial producers of hydrogen [9]. The hydrogen market comprises two main players: captive producers which produce hydrogen for their direct customer or their own use and by-product hydrogen resulting from chemical processes. In developed or technological advance countries is currently identified a third player: merchant companies, which trade hydrogen. In Romania, hydrogen was consumed by mainly two industrial sectors: the refinery and the ammonia industry, which are both captive users. Besides the two categories of manufacturers, petrochemical and agrochemical, there is a third category chlor-alkali industry where hydrogen comes as by-product during brine electrolysis. Hydrogen can convert chlor-alkali technology into a more efficient one, may save a certain amount of energy. Till at this time hydrogen produced by Romanian industry is not sold, it is used internally or is captive. The hydrogen produced by electrolysis of brine is most suitable for applications in fuel cells. The main producers of hydrogen by this process are: *Oltchim* S.A., Rm. Valcea, and *Chimcomplex* S.A., Borzesti [10]. The former *Oltchim* S.A. had a tentative to install in Romania a fuel cell stack with a capacity of 2.0 MW. The idea of the project was to produce clean electricity by using the hydrogen co-generated in the electrolytic process for chlor-alkali production. Being one of the first installations in this field, its cost was higher than other more consolidated technologies, thus requesting the appropriate supporting frame. The majority of the cost structure comes from two major items: the fuel cell capital, and maintenance cost, and the industrial balance of plant, still too complex to accommodate the fuel cells operating requirements [11]. *Chimcomplex* S.A. recently has acquisitioned new equipment for brine electrolysis, which significantly increases the capacity of industrial production of hydrogen.

A hydrogen strategy for Romania must take into account the geopolitical factors that affect it, its state of economic development, and the social awareness for a hydrogen economy. The hydrogen roadmap would set the direction for the changes, could feature future stages in logical succession, and its timeframe. As a general remark, it is observed that countries that have difficulties in promoting and implementing the hydrogen technologies have ambitious plans rather than realistic. Adequate funding and decision of experts are key elements for the success of such a hydrogen research program. Taking in consideration the experience gained in the United States, Japan, South Korea, Canada, and at the level of UE, the authors repeat, not for the first time, main items necessary to promote hydrogen and fuel cell:

- Clear statement intention of the authorities, translated into a coherent and realistic scheme, based on germination of experts,
- Formation of hydrogen communities or at least of initiative groups,
- Inter-ministerial consultations with entities responsible for research, development and demonstration policy, and the finding of the funding and financial schemes of the implementation projects,
- Promotion of research, development, and demonstration projects in close cooperation with the international exchange of knowledge, and
- Targeting the transition to commercial scale for generating social and economic benefits.

The industrial processes that may be used to obtain or to use sustainable hydrogen in Romania include:

- Steam methane reforming with carbon capture and sequestration,
- Hydrogen from biomass and bio-hydrogen,
- Renewable hydrogen,

- Nuclear hydrogen,
- Hydrogen for energy storage, also very important into the context of hydrogen energy.

The order in which issues were listed is not a classification or ranking and is just coincidental. A prioritization of these directions should be made after a detailed analysis involving political and economic stakeholders, authorities, the public, and the scientific community finally. In countries with a progress in the implementation of hydrogen and fuel cell technologies is observed that such decisions were made by policy makers based on a prior expert analysis. The recommendations are considered: clear political support alongside developing partnerships with international partners investing substantial in hydrogen technology, increasing the level of expenditure (investments) into research, development, and deployment of hydrogen technology, and focus on the niche markets in close connection with national opportunities.

6.3 Aspects about the Implementation of a Hydrogen Infrastructure Starting with the Hydrogen Underground Storage in Romania

The hydrogen storage at large scale is expected to sustain the integration of intermittent renewable energy sources in the energy system. Large quantities of hydrogen can be produced from renewable electricity through the electrolysis of water. The hydrogen storage in caverns is a suitable candidate for dynamic peak load energy storage and hydrogen can be released within an adequate period of time. The caverns are highly impermeable for gases and are virtually leak proof. The cushion gas requirements for the hydrogen storage are large and depend on the minimum possible cavern pressure, but are defined as economic inventory [12].

The literature reported on the analysis work on physical, chemical, environmental, and energy issues, also numerous case studies were developed and described by authors in other papers [13]. Early open literature references about hydrogen underground storage were made in 1976 and 1979. A major study on the same topic was conducted by the Institute of Gas Technology, USA. A few years later were discussed issues related to technical and economic aspects, a scenario about hydrogen underground storage in salt caverns in connection with the use of renewable energy for balancing out tidal energy fluctuations. Others disclosed a conceptual design for compressed hydrogen storage in mined caverns, that is, in excavated tunnel-shaped caverns. At that time hydrogen underground storage appeared to be a promising solution to the problem of large scale energy storage. Despite the principle idea of large-scale hydrogen underground storage in salt caverns is already three decades old, much work remains to be done to put real systems into operation and demonstrate their effect. Having disappeared from literature for about 15 years, the idea of hydrogen underground storage reappears in 2000. Diverse sources disclose relevant comparisons between hydrogen underground storage and high-pressure over-ground storage, the potential of large scale hydrogen storage in the UK, the USA, Denmark, Germany, and Russia. Also one study has reported about a possible integration of large-scale hydrogen underground storage into European energy infrastructure. In our days, a number of industrial hydrogen underground storage applications are in operation. Chemical industry is operating hydrogen salt caverns in Clemens and Moss Bluff in the U.S., as well as in Teesside in the UK where hydrogen is stored in three small and shallow caverns, using pumping of brine into aboveground brine ponds instead of differential pressure for hydrogen storage underground. In those cases, the stored hydrogen is used in chemical industry. Also, in the past, hydrogen has successfully been stored underground in France, Germany, and former Czechoslovakia, as pure hydrogen for aerospace industry needs in Russia, or as town gas: a gas mixture including hydrogen (40–60%), carbon monoxide, methane, and volatile hydrocarbons.

This work has originated from a European assessment project by the name of *HyUnder*. The project was supported by the FCH JU (Fuel Cell and Hydrogen Joint Undertaking, grant no. 303417) and has set out to reveal more about the storage potentials, relevant salt and other relevant underground energy storage geologies, process technology and cavern operating conditions, potential business models and relevant energy markets for the use of large-scale hydrogen underground storage in Europe. Romania is one out of six regions serving as prototypical energy market with sufficient salt structures. In the project, the system boundaries are defined such that the analyzed cavern plant includes electrolysis, compression prior to the underground salt cavern as hydrogen storage and all topside equipment (i.e., hydrogen drying, purification, compression for trailer filling, re-electrification unit, and NG grid injection unit).

A future hydrogen storage infrastructure may take years lead time for its development. The early scientific, technical, and business analysis are needed in time, involving representatives of the relevant communities to share their know how in an attempt to develop realistic plans that may help to better understand the role of hydrogen in Romania's future renewable energy-based energy system. The best approach to develop this pathway is by developing the transition using both a long-term and short-term approach. Romania has a long tradition of salt extractions. Today, Romania operates several active mines, but some older mines have also been closed. Many of these sites may have the potential to be used for hydrogen storage to form part of a wider hydrogen infrastructure.

The transition of the hydrogen wider integration into the energy market will start with incipient hydrogen communities and early niche markets, for both stationary use and mobility. The hydrogen economy will assume not only the existence but also the continuity of centrally organized energy systems, the introduction of hydrogen into the energy systems, the development of a hydrogen distribution grid for transferring hydrogen from the locations of production to the sites of consumption. As part of this infrastructure, energy and hence hydrogen storage would play an integral and important role. The scientific literature assumes that the development of a hydrogen pipeline network for transport and distribution could take as long as 60 years and the centralized infrastructure could cost half of a decentralized hydrogen infrastructure [14]. The hydrogen could be supplied both for vehicles via refueling stations and to the industrial and residential sectors for electricity and heat requirements. The hydrogen economy, secure and cheap hydrogen infrastructures will be needed in the future. The balance between the use of hydrogen onsite and storage of hydrogen from distribution network will be dictated by both the complementarity of supply and demand as well as possibly regulations and geographic conditions. The future energy system will be developed with two energy provision networks in place, one for electricity and one for hydrogen, using power-to-gas technology or something derived from it (e.g., synthetic methane gas through methanation with CO_2). The hydrogen network system can be designed similar to electric transmission grids or natural gas pipeline network. In our view, the hydrogen network would comprise two subsystems: one for transport and one for distribution, the first one include a transmission ring, Figure 6.1. The potential sites where is possible to locate the large-scale hydrogen underground storage facilities in Romania must intersect the hydrogen transition ring. The storage facilities will be well connected to the national hydrogen network and will respond to the local, national, and also international requirements. Details about those potential sites, four by number, is described in the next paragraphs. The infrastructure include: hydrogen production plants, especially electrolyzers with renewable electricity; supply stations; maintenance services; connections with renewable and electrical grid; fuel cell plants; control and operating centers; stations for import-export; small-scale (aboveground or buried) storage facilities, etc.

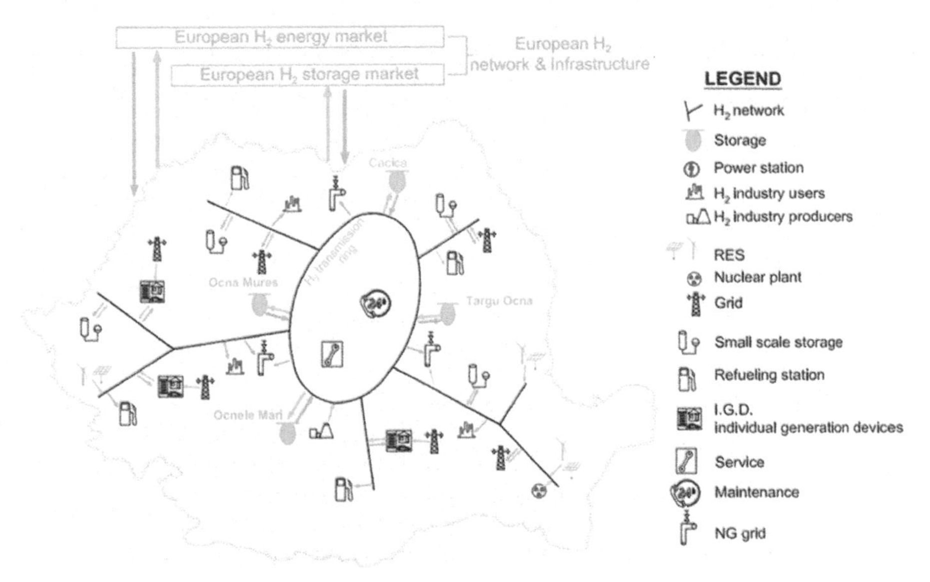

Figure 6.1 Scenario for a hydrogen infrastructure in Romania.

One possible future part for this infrastructure will be hydrogen production with nuclear energy, Table 6.2.

The national hydrogen network also must be connected to the European market; national operator(s) can export or import hydrogen. It is imperative to understand that hydrogen can be exported and imported both as energy vector and stored energy. In the future, the sale of energy storage can become a distinct market. The context indices that the large-scale hydrogen underground storage mast have a multifunctional role and the benefits will be comprehensive.

Reiterating the opinion of some authors cited in this work, this can be considered a brief description of what is pursued by visionaries who believe in a hydrogen-based future.

Romania has a long tradition in the cavern development and operation, nowadays, there are seven active sites, both mines and caverns, these locations can be evaluated from them potential for hydrogen storage. The identification of the most promising site(s) for early hydrogen storage, specifically from a market perspective, is a difficult work. In them studies, the authors decided to consider only sites where salt caverns and necessary infrastructure such as a standard procedure for brine disposal already exist. It is much easier to decide the development of the existing fields by conversion of existing caverns or adding new ones for hydrogen storage. The European project *HyUnder* marks the beginning of Romanian studies related to the possibilities

Table 6.2 The elements for a national hydrogen infrastructure

No.	Element	Brief Descriptions
1	Hydrogen pipeline transport and distribution	Used to transport hydrogen from the point of production to the point of demand. The technology for hydrogen pipeline transport is proven and the transport costs are somewhat above those for natural gas pipelines.
2	Transmission ring	As integral part of hydrogen pipeline transport contributes to system flexibility and connectivity.
3	Large storage facilities (underground)	This study has only addressed salt caverns, in principle, other geological formations may also be used.
4	Connections	Include all direct and indirect contacts with hydrogen producers and users: industry, RES, natural gas and electric grid, refueling stations, etc.
5	Hydrogen refueling stations	Storage or fueling stations for hydrogen are usually located along a road or hydrogen highway, or at home as part of the individual generation device.
6	Small scale storage	Hydrogen can be stored above ground in a gas holder largely for balancing (making sure pipes can be operated within a safe range of pressures), not long-term storage.
7	Hydrogen individual generation devices	In dedicated electrolyzers, able to produce hydrogen for individual houses and cars, these devices will be able to produce hydrogen when, for example, electricity price is low or to be connected to a neuronal network that can order hydrogen production conditioned by algorithms.
8	Intersections with European infrastructures	National infrastructure will have connections with neighboring countries as part of the European network and infrastructure. There is the opportunity to participate in European hydrogen energy and storage markets.
9	Services and maintenance	These are necessary elements for the operation of technical systems.
10	Nuclear hydrogen	The utilization of nuclear energy for large-scale hydrogen production is believed to have a key role in a sustainable energy future in Romania. Co-generation of both electricity and hydrogen from nuclear plants may become increasingly attractive. The nuclear hydrogen production mainly includes electrolysis and thermochemical cycles.
11	Hydrogen producers and users	These two elements create demand and supply on the hydrogen market.

of hydrogen underground storage, which aimed to acquire information about underground storage of hydrogen and to achieve data about the potential of the renewable energy and energy infrastructure. It is useful to understand the potential relevance and scale of underground storage of hydrogen as a mean for large-scale storage of intermittent renewable electricity in the context of the expected development of the Romanian power sector until 2050; to identify the most promising sites for potential hydrogen caverns taking into account the specific geological conditions in Romania as well as other relevant location factors; and to analyze the economic attractiveness and potential business case for underground hydrogen storage with respect to the different end-use applications of hydrogen (mobility, industry, admixture to the natural gas grid, or re-electrification). Four sites have been identified in order to be analyzed if they are suitable hydrogen storage: Ocna Mures, Targu Ocna, Ocnele Mari, and Cacica, Table 6.3. In these locations, hydrogen can be produced by water electrolysis in case of renewable energy surplus and when, undesirable situations, demand are less than production. Into this manner important quantities of energy can be saved and stored.

According with the author's criteria, which refer to the evaluation of locations, there are some general requirements largely accepted: good geological conditions, cavern field in conservation, brine consumption by the local industry or environmental conditions for discharge were identified. These criteria could be limited even if this industry is very strong in that specific region, because of the long-term contracts (up to 10 years). This leads to a saturated market, which does not need additional brine. The geological conditions are assumed to remain constant for this period and also, the feasibility criteria are only relevant for the time of construction, 2025. The possible contamination and leakage are well described in literature, special attention must be paid to the proof of tightness for H_2 (synthetic seals, interface salt-cementation-steel) and adaptation of steel components (proof of resistance against hydrogen embrittlement, use of steel suitable for hydrogen). Prototypical caverns are operated at pressures between 6 MPa and 180 MPa and a cavern size of 500,000 m^3 has been considered as reference. There is a need to respect the cushion gas, which remains in the caverns at minimum pressure, to keep the cavern stable. The gas remained is a part of capital expenditure and cannot be considered and used as a working gas. The working gas is difference between the gas content at maximum and minimum pressure. Taking into consideration the physical and chemical parameters (volume, density, compressibility factor, etc.), this means 5,000 tons H_2.

Table 6.3 The potential locations for hydrogen underground storage in Romania

Region	Comments	Site	Comments on Individual Sites
Center	• located in the central part of Romania • close to urban centers such as Cluj-Napoca, Brasov, Sibiu	Ocna Mures	• Good geological conditions • Cavern field in conservation • Brine operating infrastructure available • Uncertain demand for brine, a single customer for brine
East	• located in the eastern part of Romania • close to urban centers as Bacau, Piatra Neamt, Brasov	Targu Ocna	• Good geological conditions • Cavern field in operation • Brine operating infrastructure available • Brine consumption by local industry
South	• located in the south of Romania • close to urban centers as Pitesti, Craiova, Sibiu • good connections with Bucharest	Ocnele Mari	• Good geological conditions • Cavern field in operation • Brine operating infrastructure available • Brine consumption by local industry • Operator is interested in the Case Study • Romanian National Hydrogen & Fuel Cell Center in close proximity
North	• located in the northern part of Romania • difficult connection to major cities	Cacica	• Good geological conditions • Lack of large caverns, and implicitly appropriate infrastructure

The potential use of large underground salt caverns for the storage of hydrogen as an energy carrier in Romania is dictated by the roles of the different industry sectors. Though the general interest of each sector follows similar structures, the views or interests of individual companies will differ from each other. The general interest expressed in this work has been compiled by authors and has been reviewed by the individual Romanian stakeholders. The results do not necessarily express the views of any one singular contributing company, but instead the voice of a group of diversely acting specialists, all individuals being experts in their specific application area of hydrogen energy or the use of underground caverns for other industrial purposes in general. The role of the hydrogen to balance the electricity supply system is no new item, especially considering the possibility of energy storage in the context of renewable electricity. An official statement by Romanian authorities or regulators is not possible because the national strategy has not been defined, but numerous opinions of individuals from both public and private entities underline the idea that renewables and hydrogen will contribute to large-scale energy storage, without any single individual or group providing any time horizon. The possible role of hydrogen as an energy carrier for different industry sectors in Romania can follow next directions: fuel for transport, like "premium" segment due to willingness to pay a premium for green hydrogen, chemical raw material for industry (refineries and ammonia), injection in natural gas grid, or re-electrification and load management in the electricity grid.

Regarding utilization of hydrogen storage for mobility, we mention that Romania does not have a fleet of fuel cell electrical vehicles with hydrogen (FCEVs), except two versions of mobile hybrid electric-hydrogen platform designed, build, and used for experiments within various projects at the National Center for Hydrogen and Fuel Cell. Romania has introduced incentives for electrical vehicles in its legislation, which has, however, not been fully enabled. New policies were adopted to promote clean and energy efficient road transport vehicles, hybrid and electric, in April 2011. The policies have been implemented in two directions: grants for purchase and a scrap-page scheme for public administrations or institutions, individuals, NGOs or SMEs. Therefore, there are premises for alignment at EU policies in the field of automotive and environmental protection in the next years.

A European study mentions a FCEVs penetration of the passenger car fleet between 5% and 50% in UE by 2050 [15]. Translated to the Romanian passenger car fleet of 6.4 M vehicles in 2050, this corresponds to 0.32 M (5% penetrations) to 1.6 M (25% penetrations) FCEVs. Following the same trend as in the analysis given above for 2025 is proposed less than 0.1% FCEVs penetration scenario, 0.005 M vehicles. The authors reached these figures taking into account the previously mentioned references and market-specific aspects of Romania: number of new cars, the fleet age, trademarks weight, prices, etc. [30]. According to the same publication about the power trains portfolio for Europe and hydrogen demand by the mobility sector in Germany, the authors assumed the following scenarios: 11,000 km/14,000 km annual mileage per year with two options regarding the consumption of hydrogen 0.95 kg H_2/100 km and 0.54 kg H_2/100 km in 2025 and the same options regarding the consumption of hydrogen but with two limits of penetration, as denoted above, in 2050. Taking into account the previous assumptions (number of cars, kilometers per year and consumption), the authors calculated for automotive application hydrogen consumption in Romania between 297 tons/year in 2025 to 211,445 tons/year in 2050. Table 6.4 also indicates scenarios for hydrogen consumption by the automotive sector. To determine the expected sales quantity of hydrogen as vehicle fuel, the number of cars could be a good indication. By defining a certain radius around a potential location and comparing the number of cars registered today, Table 6.5 could be used for an extrapolation of a future hydrogen demand for FCEVs. There was assumed to have two delivery radiuses for early caverns, 100 km and 200 km.

For chemical or other industry, producing and consuming hydrogen already at very large scale, and in command of all necessary technologies and processes, hydrogen is solely treated as a commodity. Its production,

Table 6.4 The hydrogen demand for transport around a potential underground storage sites

Site	Hydrogen Consumption (tons/year)					
Year (Penetration)	2025 (<1%)		2050 (25%)		2050 (5%)	
Radius Around Site	200 km	100 km	200 km	100 km	200 km	100 km
Scenario A: 11,000 km per year and 0.54 kg H_2/100 km						
Ocna Mures	142.00	47.00	45,619.00	15,206.00	9,123.00	3,041.00
Targu Ocna	169.00	41.00	54,172.00	13,305.00	10,834.00	2,661.00
Ocnele Mari	201.00	44.00	64,627.00	14,256.00	12,925.00	2,851.00
Cacica	98.00	35.00	31,363.00	11,404.00	6,272.00	2,280.00
Scenario B: 11,000 km per year and 0.95 kg H_2/100 km						
Ocna Mures	250.00	83.00	80,256.00	26,752.00	16,051.00	5,350.00
Targu Ocna	297.00	73.00	95,304.00	23,408.00	19,060.00	4,681.00
Ocnele Mari	355.00	78.00	113,696.00	25,080.00	22,739.00	5,016.00
Cacica	172.00	62.00	55,176.00	20,064.00	11,035.00	4,012.00
Scenario C: 14,000 km per year and 0.54 kg H_2/100 km						
Ocna Mures	181.00	60.00	58,060.00	19,353.00	11,612.00	3,870.00
Targu Ocna	215.00	52.00	68,947.00	16,934.00	13,789.00	3,386.00
Ocnele Mari	257.00	56.00	82,252.00	18,144.00	16,450.00	3,628.00
Cacica	124.00	45.00	39,916.00	14,515.00	7,983.00	2,903.00
Scenario D: 14,000 km per year and 0.95 kg H_2/100 km						
Ocna Mures	319.00	106.00	102,144.00	34,048.00	20,428.00	6,809.00
Targu Ocna	379.00	93.00	121,296.00	29,792.00	24,259.00	5,958.00
Ocnele Mari	452.00	99.00	144,704.00	31,920.00	28,940.00	6,384.00
Cacica	219.00	79.00	70,224.00	25,536.00	14,044.00	5,107.00

Table 6.5 Numbers of car and percentages, defining a certain radius around a potential location

	200 km Radius Around Site		100 km Radius Around Site	
Site	Number	Percentage (%)	Number	Percentage (%)
Ocna Mures	2,548,730	48	849,590	16
Targu Ocna	3,026,620	57	743,370	14
Ocnele Mari	3,610,700	68	796,480	15
Cacica	1,752,250	33	637,180	12

Number of cars today in Romania [16]: 5,309,856 (100%).

distribution, and end-use are driven by costs in the first place. The reduction of GHG emissions are important goals for the hydrogen consuming industry, driven by tightening public constraints and regulations (e.g., CO_2 certificates). There is still a distant future until Romanian industry will start to investigate potential synergies by offering their know-how and existing infrastructures to utilize economic synergies for reducing infrastructure costs on one side and to introduce "green hydrogen" on the other side.

Romania accounts 1.8% of total European petroleum refining capacity. The official oil needed prognosis for Romanian is about 11.0 million tons for the refinery sector in 2020. This amount will be slightly higher than 2011and comparable with 2009. The hydrogen demand from refineries might further increase in the long term (2050), the forecast is very dependent on the availability of inexpensive crude oil and the demand for petroleum products. There are no signs for new investments capacities except modernization, this determines the authors to assume that the refining capacity will remain constant, 15.0 million tons in 2012. According with literature, for this size of refining capacity, a typical hydrogen requirement will be about 75,000–120,000 tons/year. Another hydrogen-consuming industry is the one for ammonia and fertilizer production. According to European Fertilizer Manufacturers Association, a relative small increase of the consumption is expected for the next decades. In Romania, ammonia is produced at six plants and at this moment the nitrogen production is estimated at 1,100 thousand metric tons, stoichiometric, it is calculated a necessary of hydrogen about 235,000 tons/year for this agrochemical industry needs.

Romanian natural gas industry is a mature one. In this industry, 41,391 persons are employed and approximately 3,122,000 customers are served including domestic and non-domestic ones (industrial, commercial and other). The total length of pipelines is 53,666 km, from where the National Gas Transmission Company operates a network of 13,000 km. The natural gas sales by sector have in the previous years next values: industry 44.8% (67.6 TWh), residential 27.9% (42.0 TWh), and power plants 23.4% (35.3 TWh) from a total of 150.8 TWh, with a growth of 3% compared to 2010.

There are numerous technical solutions to develop potential future business cases: admixture of hydrogen to the existing natural gas transport grid (2%, 5% or 10%); Power-to-Gas concept, that is, methanation of electrolytically produced "green hydrogen" with CO_2 from several resources (power plants, biogas plants, other industry and extraction from air) or only lately; or the conversion of natural gas sub-grids to 100% hydrogen use. Until now, no decision has been taken by the gas industry, which of the options will be selected for which application. Different concepts may coexist and that different options may be preferred. It is reasonable to assume that, taking into account the technical aspects, the injection of hydrogen for domestic users, heat and power, it is feasible in the future. The residential represents 27.9% and power plants 23.4%, this means that 51.3% of total natural gas is burned for heat and power, ratio which can be mixed with hydrogen. According to the most recent data available, Romania has consumed about 13.5 billion cubic meters natural gas in 2012. Although in recent years consumption was relatively constant, for 2025 this amount is forecasted to increase by 11%. The amount of hydrogen to be injected into the natural gas transmission pipelines is estimated to be 12,470 tons/year (2%); 31,160 tons/year (5%); and 62,390 tons/year (10%), and potentially be increased with the aforementioned rate.

The hydrogen applications in order to balance the electricity supply system is also possible, scientists discuss about this, especially considering its role for energy storage and here again to balance renewable electricity. The power sector agrees that hydrogen or methane, if is discussed about entire power-to-gas concept, will be stored at a large scale to globally contribute to electricity-energy storage and economic synergies with other hydrogen consuming industry sectors will be possible. The hydrogen storage could then both contribute to level out the renewable electricity surplus at supply side and electricity shortages at user side, as well as help to solve local grid congestion challenges. The future business cases would then be driven by "volatility" vs. "spark spread" (no energy transformation).

The Romanian electricity production of renewable hydrogen is considered for a scenario where renewable, specifically wind energy, will reach an installed power of 4,000 MW by 2025, and 6,000 MW by 2050, respectively, the number of full loads hours being 2,350 per year. From this a portion of 5–10% is considered to be in excess and will be needed to be stored as hydrogen. The hydrogen produced from excess of renewable electricity must be consumed for power needs using fuel cell applications, both mobile and stationary. Taking into account the aforementioned amount of electrical power, electrolyzers efficiency and lower heating values, it results that production will be between 9,600 H_2 tons/year and 19,200 tons/year by 2025, and 14,400 H_2 tons/year and 28,800 H_2 tons/year by 2050, respectively.

The electricity production from hydrogen will probably take place at sites with strong grid connections (e.g., sites of today's existing plants) or at least at sites with transformer station (or electrical substations) of Transmitter System Operator (TSO) of 220–400 kV. The locations for hydrogen underground storage in salt caverns are in the vicinity of an electricity distribution network with transformer station of 110 kV. Around of all four cavern locations, there are no important power plant capacities, the authors do agree that the distance to the transformer station of TSO (220–400 kV) and the distribution stations (110 kV) will play an important role in the future development of hydrogen-to-power infrastructure, Table 6.6.

The identification of the sites for storing hydrogen into caverns was done considering the salt caverns infrastructure. Using aforementioned data, the author was able to present a graph of the possible hydrogen needs, Figure 6.2. It is good to understand that those numbers only give a rough estimate of the possible future hydrogen demand in Romania.

The large-scale hydrogen storage facilities are obviously dependent on the hydrogen infrastructure build-up in Romania. National, multinational, or European hydrogen infrastructures imply both large- and small-scale hydrogen storage. The best options for the large quantities storage of hydrogen over long periods of time are salt caverns. The sectors indicated in the earlier paragraphs and Figure 6.2 have shown a general interest for an analysis of the large- scale hydrogen storage in salt caverns. The chemical industry will become interested to participate in this market if the cheap surplus of hydrogen quantities would become available.

The hydrogen stationary applications and the fuel cell technologies in the energy sector, both electricity and natural gas, could solve the problem of intermittent transmission by seasonal energy storage with hydrogen energy at

Table 6.6 Distances between selected sites and transformer stations of TSO

Site	Station: 220–400 kV	Station: 110 kV
Ocna Mures	50 km	<10 km
	25 km	
Targu Ocna	30 km	<10 km
	25 km	
Ocnele Mari	10 km	<10 km
Cacica	30 km	<10 km

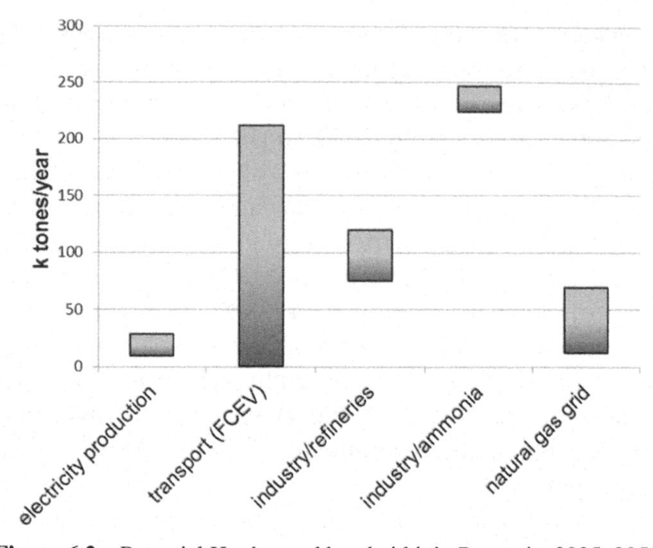

Figure 6.2 Potential H_2 demand bandwidth in Romania, 2025–2050.

large scale. In the short and medium term, Romania's energy strategy foresees to provide an increased gas storage capacity of up to 50%, an increasing oil storage capacity to fulfil 67.5 days of annual consumption and the construction of a 1,000 MW new pumped hydro storage plant. These latter figures (6.3, 6.4, 6.5, and 6.6) are reasons to appreciate the potential of hydrogen underground storage in the Romanian energy sector.

It is useful to note that the hydrogen storage should not be a purpose itself; it represents only a part of the total costs of the hydrogen value chain. The storage represents around 5% of the total natural gas price paid by residential clients into a mature market. At the incipient stage, the main roles of hydrogen storage can be considered similar as those for natural gas storage: security of supply, system flexibility, production and transmission optimization, infrastructure sizing, market development or arbitrage.

At this moment, the costs of hydrogen, produced by water electrolysis using renewables, appears to be far more expensive than hydrogen produced by steam methane reforming. One solution is to store and use hydrogen as energy carrier, which does not necessarily mean to produce continuously [17]. The hydrogen is just one part of the energy solution; therefore the suggested approach of a country-wide large-scale hydrogen energy delivery and storage system needs more dynamic analysis. There are not all instruments in place to create a map of the role of energy storage in general and the hydrogen underground storage into salt caverns in particular.

6.4 The Hydrogen and Fuel Cell Research and Development Projects in Romania

The present paragraphs are based on study of the authors about a "snapshot" of research, development, and innovation into the field of hydrogen and fuel cell where we identified the programs and projects in which the Romanian hydrogen community is active. The author's enumeration of the hydrogen and fuel cell research projects is followed by brief comments, without pretending to be a critical and multidisciplinary analysis. The work aims to present a statement of research projects, development, and innovation in Romania.

The Romanian hydrogen actors in the field of research and development of hydrogen and fuel cell have been conducting investigations on a number of issues related to this topic. These efforts concern the generation and storage of hydrogen, fuel cell construction, and their electric and transport properties. In Romania, most public funding on research and development is allocated by the research authorities. It is necessary to develop a global perspective and to follow the latest trends in energy. The development of new solutions of energy-related problems will also depend on the co-operation of numerous regional or/and international stakeholders.

Starting with 2000 till 2005, the projects that referred to hydrogen and fuel cell were financed mainly through national programs of research and development entitled: MENER, MATNAN, CERES, *Sectorial Plans* of R&D, and *Nucleu* Programs. An investigation in existing databases an inventory of the research and development programs referring hydrogen and fuel cell was conducted by authors. In the framework called MENER, (medium, energy and resources), the hydrogen and fuel cells found their place in subprogram C (new renewable and conventional energies).

Table 6.7 reveals the projects performed in the period 2000–2005, there were 11 projects, 7 participants, and covered 5 programs; the amount of money invested was estimated by authors to be under a million of Euro.

The *Sectorial* Program was focused on: improvement of the scientific and technological capabilities of the partners in the hydrogen and fuel cells domain, determining future directions of research and development in this area, in close contact with European strategy, increasing national competitiveness on the research and development of hydrogen fuel cells. The consortium was leaded by ICSI Rm. Valcea, and the members were four other national research institutes; ICPE-CA Bucharest, ITIM Cluj Napoca, IFT Iasi, and INFM Magurele; two universities: Oil and Gas University of Ploiesti and Polytechnic University of Bucharest; and two private companies ZECASIN S.A. and ICEMENERG S.A. The project research directions was PEMFC and SOFC

Table 6.7 The hydrogen and fuel cell projects in Romania, 2000–2005

	Project	Partners	Program, Year
1	Integrated system of energy production based on hydrogen fuel cell and proton exchange membrane	INCDTCI-ICSI Rm. Valcea, INCDIE-ICPE-CA Bucharest, University of Pitesti, Technical University of Cluj Napoca	MENER, 2001
2	Conversion and energy storage technologies using hydrogen fuel cells for telecom	INCDTCI-ICSI Rm. Valcea, ITIM Cluj Napoca, INCDIE-ICPE-CA Bucharest	MENER, 2002
3	Research on geometric and thermodynamic optimization of PEM FC stack	INCDTCI-ICSI Rm. Valcea, INCDIE-ICPE-CA Bucharest	MENER, 2004
4	Fuel Cell experimental model for optimizing gas distribution and water balance	Ovidius University of Constanta	MENER, 2004
5	Ceramic Composites for SOFC-IT	INCDIE ICPE-CA Bucharest, University of Bucharest	MATNAN TECH, 2000
6	Ceramic electrodes used in SOFC cells of medium temperature	INCDIE ICPE-CA Bucharest, University of Bucharest	MATNAN TECH, 2001–2004
7	Ceramic selective membranes for oxygen used in chemical processes at high temperature	INCDIE ICPE-CA Bucharest, University of Bucharest	MATNAN TECH, 2001–2004
8	Storing hydrogen isotopes in metal compounds	INCDTCI-ICSI Rm. Valcea, INCDIE-ICPE-CA Bucharest	CERES, 2001
9	Studies on methods of hydrogen storage in soft metal alloys	INCDTCI-ICSI Rm. Valcea, INCDIE-ICPE-CA Bucharest University of Pitesti	CERES, 2004
10	Techniques and materials for hydrogen storage	INCDIE-ICPE-CA Bucharest	"Nucleu" Program, 2004–5
11	Development of an integrated research and development of hydrogen and fuel cells in Romania	INCDTCI-ICSI Rm. Valcea	Sectorial Program 2004

fuel cell, hydrogen production through reforming, hydrogen purification, hydrogen storage in metallic hydrides and biofuel.

The programs named CEEX, Research Excellence issued by Government Decision no. 367 from 28 April 2005, financed research and development projects starting with 2006. In the first year, the CEEX Program, Module 1, complex projects, financed three projects. The projects supported under the umbrella of these calls are indicated in Table 6.8. For first call, one project

Table 6.8 Projects financed within CEEX competition

	Project	Leader	Program, Subprogram
1	Research on the production, purification, hydrogen storage obtained by gasification of biomass	OVM ICCPET Bucharest	CEEX 2005 Module 1
2	Alternative low-cost multifunctional materials for polymer electrolyte fuel cells (PEMFC) operating at temperatures above 180°C	INCDIE ICPE-CA Bucharest	CEEX 2005 Module 1
3	Develop and implement new solutions to improve performance of fuel cell proton exchange membrane	INCDTCI-ICSI Rm. Valcea,	CEEX 2005 Module 1
4	Hydrogen storage materials for carrying out a thermal compressor for ultra-pure hydrogen used for fuel cell and hybrid vehicles	INCD ITIM Cluj Napoca	CEEX 2006 Module 1
5	Reducing greenhouse gas concentrations by partial replacement of oil with hydrogen in industrial installations	ICEM S.A. Bucharest	CEEX 2006 Module 1
6	Research materials and advanced equipment for generating hydrogen for use in (reconverted) transportation systems and fuel storage	University of Bucharest	CEEX 2006 Module 1
7	Advanced hydrogen storage materials for fuel cell supply	INCD ITIM Cluj Napoca	CEEX 2006 Module 1
8	Develop a system for hydrogen production at low cost by the method of proton exchange membrane electrolysis	INCDTCI-ICSI Rm. Valcea,	CEEX 2006 Module 1
9	Photo-electrolytic production of hydrogen	INCD ITIM Cluj Napoca	CEEX 2006 Module 1
10	Hybrid system for autonomous power using photovoltaic/fuel cell system.	INCDIE ICPE-CA Bucharest	CEEX 2006 Module 1
11	Heat and water management systems for PEM fuel cells	INCDTCI-ICSI Rm. Valcea	CEEX 2006 Module 1
12	Microstructure of αAl2O3-ZrO2 micro and nano-metric systems doped with rare earths for efficient composite materials (solid electrolyte in intermediate temperature fuel cells SOFC)	INCDFM Bucharest	CEEX 2006 Module 1
13	Research on intelligent systems and environmentally friendly surface-based transport and fuel cell electric drive	SC Setko Impex srl	CEEX 2006 Module 1
14	Next-generation solid electrolyte super-acid fuel cell (SAFC), for operation at temperatures below 200°C	University of Bucharest	CEEX 2006 Module 1

(*Continued*)

Table 6.8 Continued

	Project	Leader	Program, Subprogram
15	Promoting the integration of Romanian research teams of hydrogen technologies and fuel cell research programs developed under the auspices of the European Technology Platform for Hydrogen	"Ovidius" Universiry Constanta	CEEX 2006 Module 3
16	Develop national and European partnerships in the field of materials for hydrogen storage units, in line with the objectives of the European Technology Platform for Hydrogen and Fuel Cells	INCDIE ICPE-CA Bucharest	CEEX 2006 Module 3

was entitled "Low cost multifunctional alternative materials for polymer electrolyte fuel cells which operate at temperature above 180 degrees Celsius", proposed by a consortium led by ICPE-CA Bucharest. Other project, about developing and implementing new solutions to improve performance of fuel cells was conducted by a consortium led by INCDTCI-ICSI Rm. Valcea. The third one was focused on gasification of biomass and hydrogen production and was coordinated by OVM ICCPET. In the next call, from 2006, the number of projects covering the hydrogen energy domain increased significantly, being able to enumerate eleven projects included in Module 1, complex projects, and another two included in Module 3, projects related to promoting participation in European and international programs. The projects consist in studies of the materials involved in systems that use hydrogen and fuel cells, storage materials, equipment for hydrogen or fuel cells for intermediate temperatures; therefore, an important category of projects refers to the use and production of hydrogen. In our works, we also underlined the proposal dedicated to young scientists, named "Hydrogen, a new energy vector. Applications in tertiary sector," even if it was not accepted for funding, the proposal indicated the interest of young scientists for this sector at that moment.

These efforts have created prerequisites for germination of a hydrogen community. Starting with this concatenation, next calls has led to a coagulation and interconnection with the industry. It is true that this approach between research and industry was not a fruitful story yet.

The National Plan for Research, Development, and Innovation (PNCDI II) was lunched by public authorities responsible for funding research development for period 2007–2013. The projects were financed, based on a

Table 6.9 Projects financed within PNCDI II competition

	Project	Leader	Program, Subprogram
1	Photo-catalyst for hydrogen production and fuel from biomass and wastewater	University of Bucharest	PNCDI II 2-Energy 2007
2	The implementation of clean energy technologies by developing a heat engine based on hydrogen absorbing metal alloys using solar energy or residual energy	INCD ITIM Cluj Napoca	PNCDI II 2-Energy 2007
3	Hydrogen from bio-renewable resources	Petrol&Gas University of Pliesti	PNCDI II 2-Energy 2007
4	Development of an integrate production of hydrogen and fertilizer for the soil by use of biomass and residues	INCDTCI-ICSI Rm. Valcea,	PNCDI II 2-Energy 2007
5	Research and development of membrane reactor for the production of ultrapure hydrogen used in fuel cell power	INCDTCI-ICSI Rm. Valcea,	PNCDI II 2-Energy 2007
6	Develop a membrane-electro-catalytic system for the generation, separation, and purification of hydrogen in different environments, applicable to heat engines	CCMMM SA Bucharest	PNCDI II 2-Energy 2007
7	Green energy without oil-based: multiple vectors renewable hydrogen production, remote sensors, and designed for. SMEs critical irrigation, anti-desertification, etc.	IPA SA Bucharest	PNCDI II 2-Energy 2007
8	Regenerative electrolyzer-fuel cell energy converter, architecture design and implementation	INCDTCI-ICSI Rm. Valcea,	PNCDI II 2-Energy 2007
9	Integrated 5 kW fuel cell module	INCDIE ICPE-CA Bucharest	PNCDI II 2-Energy 2007
10	Nano-hybrid systems for electro-catalysis in renewable fuel cells	Politehnica University of Bucharest	PNCDI II 2-Energy 2007
11	Advanced ceramics components of the intermediate temperature fuel cells	ICF "Ilie Murgulescu" Bucharest	PNCDI II 7-Materials 2007
12	Advanced ceramic Nano-composites for a new generation of solid electrolyte fuel cells, of medium temperature (IT-SOFC)	INCD IFT Iasi	PNCDI II 2-Energy 2007
13	Innovative system for power using high temperature PEM fuel cells and hydrogen produced by reforming of acetic acid	INCDTCI-ICSI Rm. Valcea,	PNCDI II 2-Energy 2008

(*Continued*)

Table 6.9 Continued

	Project	Leader	Program, Subprogram
14	Composites for efficient storage of hydrogen at ambient temperature, clean energy source for fuel cells	INCD ITIM Cluj Napoca	PNCDI II 2-Energy 2008
15	Photo-catalytic hydrogen production using solar energy by using industrial waste sulfur (H_2S, SO_2)	INCEMC Timisoara	PNCDI II 2-Energy 2008
16	Technology and test device for increasing the efficiency of fuel cells operating with low concentrate fuels	Politehnica University of Bucharest	PNCDI II 2-Energy 2008
17	Ecological technology for gasification with the pure hydrogen production	Politehnica University of Bucharest	PNCDI II 3-Environ. 2008
18	Environmental impact analysis in the context of the widespread use of hydrogen-based technologies	INCDTCI-ICSI Rm. Valcea,	PNCDI II 3-Environ. 2008
19	Reduction of sulfur and ash from coal burning by using on hydrogen injection technology	Politehnica University of Bucharest	PNCDI II 3-Environ. 2008
20	New boride and nanostructured hydrides for hydrogen storage	INCD IFT Iasi	PNCDI II 7-Materials 2008
21	New complex hydrides for hydrogen storage in automotive applications	INCDFM Bucharest	PNCDI II 7-Materials 2008
22	New chemical systems based on nano-crystalline frameworks and porous architectures for Intermediate Temperature Solid Oxide Fuel Cells(IT-SOFC)operating with biogas	Ilie Murgulescu Institute of Physical Chemistry, Romanian Academy	PNCDI II 2-Energy 2011
23	Simultaneous bio-hydrogen production and wastewater treatment by selectively enriched anaerobic mixed microbial consortium	Polytechnic University of Timisoara	PNCDI II 3-Environ. 2011
24	Combustion chamber with flame turbulent experimental model, with hydrogen-enriched natural gas	National Research and Development Institute for Gas Turbines COMOTI	PNCDI II 3-Environ. 2013
25	Bioenergy generators: Design of new electrocatalysts for PEMFC (fuel cells) based bioethanol for portable applications	University of Bucharest	PNCDI II 2-Energy 2013
26	Power stationary sources with fuel cell power for bio-organic farming in greenhouses	University of Bucharest	PNCDI II 2-Energy 2013

competition. The program financed 26 projects, the projects were created as a consortium led by research institutes, universities, and research centers.

The proposed and financed projects represented a multidisciplinary approach for identifying the best scientific and technological solutions to reach the desired objectives. As a consequence of the program, the projects led to joint efforts of major researchers in the hydrogen energy and fuel cells. These calls were a bridge between national and European programs. From practical point of view was a copy rather than an implementation of the EU model, which requires some rearrangements specific to local conditions. It was a useful public financial effort, but needed more improvements. The inconsistencies can be explained considering the fragmented research that exists in Europe not just the specific nationwide problems.

The projects were focused on research about PEMFC and SOFC applications, use of new materials and integration with other regenerative systems. Referring to hydrogen, there were projects that relate to production, purification, storage, and biomass multiple roles in production of hydrogen. The specialization of research groups on specific topics such as hydrogen storage in metal hydrides, PEM fuel cell, SOFC fuel cells, hydrogen production by reforming, advanced ceramic materials has been observed by authors.

In 2007, 12 projects were financed involving 30 entities, state or private companies. The consortiums varied between three and five partners. Projects covered topics related to production, use and impact of hydrogen usage, fuel cells, using of fuel cells and materials for fuel cells. In 2008, the number of projects funded decreased to 9, which can be related with a drop in project competitiveness. Politehnica University of Bucharest awarded that most projects, followed by ICSI Rm. Valcea. In the final two calls, 2011 and 2013, the number of financed project continue to decrease drastically, there were financed only 5 projects. From an administrative point of view, the projects were divided in two thematic areas: energy (subprogram 2), environment (subprogram 3) and thematic materials, processes and innovative products (subprogram 7).

The research collective from INCDTCI-ICSI Rm. Valcea, somehow more versatile, has covered more research fields: research regarding PEM type fuel cell, hydrogen production by reforming methane or biomass gasification, hydrogen storage and fuel cell integration in diverse applications. This situation is explainable by the realization of the National Center for Hydrogen and Fuel Cell at Rm. Valcea. The group from ITIM Cluj-Napoca developed projects focused on hydrogen storage in hydrides and composite materials and the institute has a long tradition in this area, in the 80's they made a car where the hydrogen was stored in hydrides. Similar research interests were found at INCD IFT Iasi and INCDFM Magurele, the difference are that the final two institutes are more focused on physics than chemistry. The SOFC research is carried on at Politehnica University of Bucharest and Chemistry and Physics Institute “Ilie Margulescu”. The using of fuel cell assembly for diverse applications is carried out at INCDIE ICPE-CA Bucharest. The research regarding materials can be observed at University of Bucharest and University of Pitesti. COMOTI institute is interested more on the utilization of hydrogen as fuel for classical combustion. The earlier mentioned are not intended to exclude contributions from other research groups, the review was made based on a public data about the scientific research projects funded with money from the state budget. Because in the national scientific research programs is not a topic or a direction for hydrogen and fuel cell the authors were forced to search the whole database, project by project using keywords followed by an analysis of the project. So, it is therefore possible either due to subjective reasons or misunderstanding that some projects to be omitted.

These competitions and calls encouraged the cooperation between various research groups, creating powerful connections. Researchers and institutes had the opportunity to know each other, to observe both scientific areas of interest co-operators and competitors, to know not only their opinions but also their misunderstandings. The calls and projects were characterized by serious delays and budget amendments, which caused the qualitative and quantitative results of the projects to be seriously altered.

Using their own investigations, the authors have estimated that Romanian research authorities have invested maximum 26 million Euros for hydrogen and fuel cell research since 2000. This includes direct financing of the projects, see Table 6.10, and investments into the Romanian National Hydrogen and Fuel Cell Center, approximated at four million of Euro. These expenses were supplemented by equipping the laboratories with new equipment, different actors using their own funds or financing through European Union programs. In general, the governmental measures enumerated earlier prevented migration of skilled labor and dissolution of research community. Regarding hydrogen and fuel cell research and development, can be observed a decreasing of interests in the previous years. The maximum numbers of projects financed by the competitive calls was between 2006 and 2008, after that the number decrease only at 2–3 par call. And all of that taking into consideration that at the UE level, by means of FCH2 JU, the number is constant and the budget is increasing.

As a final and important element of comparison, it must be noted that those, who funded the Platform for Hydrogen and Fuel Cell from Romania,

Table 6.10 Projects financing with public funds per call in Romania

Period	Program	Number of Projects	Estimated Amount (Million Euros)
2000–2004	MENER	4	0.340
	MATNAN TECM	3	0.107
	CERES,	2	0.113
	"Nucleu" Program	1	0.043
	Sectorial Plan	1	0.325
		Total: 11	Total: 0.928 (approx. 1.0)
2005–2006	CEEX 2005	3	
	CEEX 2006	13	0.5, per project*
		Total: 16	
2007–2013	PNCDI II 2007	12	
	PNCDI II 2008	9	
	PNCDI II 2011	2	0.5, per project*
	PNCDI II 2013	3	
		Total: 26	
		Total (approx.)	22.00

*initial amount, during project development the amounts were reduced due to budget rectification.

were that main actors in the implementation of national research projects in the field of hydrogen and fuel cells.

6.5 Romanian's Participations in FCH JU Calls and Projects

Starting with 2008, Romanian's research entities started the participation in FCH-JU calls with a successful funding proposal for a maximum duration of the project of 36 months. The authors have estimated that Romanian researcher groups participated in a little more than 10 proposals for FCH JU calls. As a general characteristic, the success rate was not only very small but also the number of ineligible proposals is very small, however, can be deduced that the most of these proposals have not met the required score to be selected for funding.

Two research institutes in Romania, namely National Research and Development Institute for Cryogenics and Isotopic Technologies ICSI Rm Valcea and Institute Oskar Von Miller—SC OVM-ICCPET SA, participated in projects entitled: "Assessment of the potential, the actors and relevant business cases for large scale and seasonal storage of renewable electricity by hydrogen underground storage in Europe (*HyUnder*)" and "Molten Carbonate Fuel Cell

Catalyst and Stack Component Degradation and Lifetime: Fuel Gas Contaminant Effects and Extraction Strategies (*MCFC-CONTEX*)". Activities relates to (1) a case study concerning underground storage of hydrogen, (feasibility, relevance, timelines, chances, and limitations of H_2 underground storage to facilitate renewable electricity in Europe); and (2) the investigation of poisoning mechanisms caused by alternative fuels and applications and determining precisely MCFC tolerance limits for long-term endurance, and optimizing fuel and gas cleaning to achieve tailored degrees of purification according to MCFC operating conditions and tolerance. A third entity, Pirelli & C. Eco Technology Ro S.R.L., was identified as a part of an consortium that has been selected for funding in a project called "Biogas robust processing with combined catalytic reformer and trap (*BioRobur*)", but it was not included in the final list of project partners.

In chronological order, the first project financed by FCH JU, was *MCFC-CONTEX*. The project aims to tackle the problem of degradation by trace contaminants. They investigated the potential for active CO_2 separation from power plant flue gas (generating power instead of consuming it in the process) and determination of poisoning mechanisms caused by SO_2 in the tail pipe gas, also through numerical modelling and accelerated testing. Second task was the optimizing clean-up of biogas from waste-water treatment and natural gas to achieve tailored degrees of purification according to MCFC operating requirements. There were 11 partners from 6 countries. The second one, *HyUnder* was a consortium with 12 organizations from 7 different European countries including large companies, small medium enterprises, and research institutes. This variety ensures that all the fundamental competencies are available to carry out the project, including: geology of underground formations suitable for gas storage and below ground technology, underground storage engineering along with above ground process technology.

As a general characterization, the proposals and financed projects with Romanian partners represented a multidisciplinary approach for identifying the best scientific and technological solutions to reach the desired objectives. As a consequence of this program, two Romanian partners were implied into a direct cooperation with the European team researchers in the hydrogen energy and fuel cells, and others were encouraged to do it.

Regarding Romanian presence in FCH JU, the main problem is that financial support is assured by minister of education and research, other ministers like: economy, transport, development, are not financing or co-finance hydrogen and fuel cell projects. Also, in Romania the industry seems not to be interested by the subject of hydrogen and fuel cell; despite of the fact that the renewable sectors have a lot of investment and the hydrogen can be an element to improve the efficiency and flexibility of it. In contrast, the countries with a remarkable progress into this direction the main financial resources are assured by industry and the authorities that sustain the renewable sector. As a final remark, at European level, there is the trend to finance more development and demonstration projects to the disadvantage of fundamental research. That means more actors from industry than academia must be implied in FCH JU calls.

6.5.1 National Center for Hydrogen and Fuel Cell in Romania

Regarding the importance of hydrogen problems and fuel cells to a global and European level, the European Commission has initiated the creation of European Technological Platform for Hydrogen and Fuel Cell Platform in 2004, and has invited Romania to bind to this. The first step was in March 2005, at the second Annual General Conference of HFP, in which the Romanian, represented has expressed the interest of collaboration and partnership to the European Hydrogen and Fuel Cell Co-ordination Network (HY-CO ERA-NET). Romania was included with Spain, Greece, Cyprus and Brazil, in one of the work assembly group of Mirror Group, directive by Portugal. The Romania's National Platform for Hydrogen and Fuel Cell has been launched in October 2005. In August 2006, was established the groups of action regarding the deployment of transnational activities and Romania has adhered to three of these.

Since 2005, the European Commission has reorganized the activity of research from the domains considered, primary through including the commercial partners as active participants not only in taking the decisions regarding the development strategies but also in investigation activities. Therefore, the private-public partnerships have appeared in many primary domains both for a better approach of the research policy and for a better coordination and correlation of the activities at member states level and associate countries. For the particular case of hydrogen technologies, it was created the initiative (JTI—Joint Technology Initiative), named Fuel Cell and Hydrogen Joint Undertaking (FCH JU) whose official inauguration has taken place in October 2008.

The National Research and Development Institute of Cryogenics and Isotopic Technologies ICSI Rm.Valcea was constant implied in aforementioned actions. The evolution of the research activities from the institute has carried on to its transformation into a research center in order to perform in some particular domains considered being very important as a support for the economic and social life. Due to the convergence of the Romanian institute's research strategy with the European standards, the institute has decided to develop activities that, in a direct and indirect way, are useful for the development of the operating levers in energetic and environmental purposes.

Since 2001, ICSI Rm. Valcea has coordinated the research and development projects referring to hydrogen and fuel cell technologies as a part of the alternative energy context. The experience and positive results obtained by the ICSI Rm. Valcea have driven the political decision makers over the opportunity of a potentially large investment in the field of hydrogen and fuel cells in order to develop a national research center at Rm. Valcea, like a separate and distinguished entity in this area. The new research center has benefited from the beginning by a substantially financial investment in order to build new facilities and acquisitions of instrumentation, instrumental to research unity in domain at the European level.

In 2006, a pre-feasibility study regarding the construction of a research and development hydrogen and fuel cell has been made. In 2007 was elaborated the feasibility study for that research center. Based on this study and others approvals, technical project for constructing the research center was approved, November 2007. Through "The Education and Research Minister Order" (the name of the minister at that time) since 02/10/2007 the technical-economic indicators of the investment's objective named "The National Center for Hydrogen and Fuel Cell" were approved and are assigned the financing sources, the state money. The entire financial effort has been supported from the state budget, Chapter 53.01, Fundamental Research and Research-Development, title 55.01, Internal Transfers. It is understandable the fact that the investment is a research structure in which the governmental authorities have spent many resources and as a consequence it is to be expected that the research center must hereafter be supported in order to reach the expected results. Initial estimations show that 34 people will work in this center. In 2008 the works on the building up begin.

The National Center for Hydrogen and Fuel Cell accommodates a modern building, with a pleasant and functional design, spanning a surface of approximately 1000 square meters which includes laboratory section, a seminar room, administrative offices and an experiment warehouse. The center is powered by alternative energy sources such as solar panels and a windmill. Since 2010 the research center functions at full capacity.

A good example regarding the recognition of the competence and capacities of this research center is the fact that it is affiliated as a full member in N.ERGHY group. Its membership allows the center the privilege to participate in decision making on a European level regarding strategies for using Hydrogen and fuel cells but also participating on the calls that will be launched within the "Fuel Cell and Hydrogen Joint Undertaking".

As Romania will adopt similar EU policies and will be aware of the necessity to establish priorities, the research center can offer technological, experimental, demonstration, etc. support for hydrogen and fuel cell technologies. Attached to this subject, the center has an array of experimenting and testing facilities needed for conducting research upon vast possibilities to use hydrogen as an alternative energy source, which can be offered in partnership with research organizations or industrial companies. The research center has the experience, the expertise, the facilities to support any project of this type.

Among the tools capacities we can find: 1) testing and experimenting regarding obtaining hydrogen out of renewable sources, 2) development of hydrogen supply plants, 3) developing ways of storing liquid hydrogen, 4) developing and experimenting with liquid hydrogen storing possibilities, 5) development, forming and testing on a demonstrative level of fuel cells not capable of more that 25–30 kw, and 6) the technologies simulation, projection and experimentation for hydrogen utilizations. The research center has in usage a computerized testing machine for the fuel cells that produce up to 25–30 kw and that require a hydrogen flow of under 50 Nm^3/h, also provides the necessary man power needed for operating and maintaining the machine.

In order to avoid further dissipation of financial and human resources required to undergo such research, the facility offers the possibility of accommodating research groups to undergo their activities. Moreover, the research infrastructure created is of great scientific value for Romania's contribution in Europe's effort to develop and implement hydrogen technologies.

There are several different types of projects have been proposed for development due to prior experience, regarding experimentation, technical realization, testing, and technological implementation for producing energy using renewable sources, with and without hydrogen involvement: 1) systems that use solar panels for energy, 2) personalized wind mill systems that store energy using hydrogen, 3) turning bio-waste into hydrogen, 4) feasibility studies regarding cogeneration within nuclear power plants, using hybrid processes with thermo-chemical dissociation reactions of water.

The activities of development and investigation from hydrogen technologies and cell fuels have a strong multidisciplinary character. The deployment of the hydrogen economy implies significant modifications: hydrogen production, the renewable energies usages, the co-generation from nuclear centers, systems of cell fuels, etc. From these points of view, a broad approach asks for the existence of the work groups and collectives, which have to be orientated to the concrete problems solving. This multidisciplinary character inside the domain of hydrogen energy enfaces the need for coordinated activities.

The multidisciplinary character of the Center is noticeable also from the way in which the six working sections were initially thought: 1) electrochemistry section investigates and analyses of processes used not only the ones involved in fuel cells but also in the equipment obtaining hydrogen through electrolysis, 2) mathematical simulation and modulation section, this section has dedicated itself to developing gas dynamics models, using CFD modulation, Monte-Carlo and gas dynamics, 3) the hydrogen producing section, that is dedicated to all means of hydrogen production, 4) the electronics section specializes on developing hardware and software means of controlling, optimizing and usages of the equipment, 5) fuel cell section, it is in charge of making fuel cells up to optimal standards, and 6) the test and validation section, this section manages the testing and evaluation of the integrated systems. The final section has at its disposal testing warehouse and its main objective is developing standard procedures for testing and operating fuel cells and the center's integrated systems.

The research center, even if it is a new entity, it has developed network of co-operators from different environments of investigation and also some commercial partners. The completed partnerships are the result of both new and previous activities regarding the collaboration in investigation projects initiated by ICSI Rm. Vâlcea, as host institution. Once the experimental and testing opportunities are open these are thought to be extended to a national scale by creating an enterprise group, existing partnership signals from different co-operators.

Another kind of partnership refers to cooperation with universities from Romania in order to start a training and specialization program in the field of hydrogen and fuel cell technologies or renewable energies. The center has an investigation program on a short and medium terms where are mentioned the main directions towards which research and development efforts are made. These directions were defined regarding the portfolio of projects and the results of the investigation collective from previous periods as well as the national necessities from energy sector.

The National Center for Hydrogen and Fuel Cell proposes to incorporate a significant part of the research from the hydrogen and fuel cell technologies in Romania, capitalizing the results that are acquired in present. Forward there will be enumerated some of the realizations: the hydrogen production installation through the methane conversion to a standard scale;

Figure 6.3 National Center for Hydrogen and Fuel Cell.

pre-purification and purification reactors for the hydrogen generated through reformation; hydrogen storage unit, fuel cell with low power (4,5W) and high power (25W); fuel cell testing stations or demonstrative models of the solar energy conversion through fuel cell. Bearing in mind the fact that the target is to develop competitive products from both point of view, scientific and economic.

As a result of the potential showed by The National Center for Hydrogen and Fuel Cell, there developed another series of capacities close to its activities portfolio. One first example is made up by the "CRYO-HY: The R&D infrastructure development of ICSI through the creation of a laboratory for low temperature through the energetic applications of the cryogenic fluids", co-financed by European Union and Romanian Govern, using the financial tool named "The Sector Operational Program by Economically Competitive Accession", the axis 2: "The Investigation, Technology Development and Innovation for Competition", the operation 2.1.2: "The Infrastructure Development of existing R&D and the creation of new R&D infrastructure (laboratories, investigation centers)", years 2010–2013. The performed activities in this project have followed the foundation of a complex laboratory for low temperature, which will concentrate on the liquefaction investigation and the processes of storage and transport of cryogenic fluids, hydrogen and helium, with applications in the energetic area. In the cryogenic laboratory will be developed and validated a series of technologies referring both the hydrogen liquefaction and energy. The new research laboratory will investigate the liquefaction systems using specific cryostats, equipment for hydrogen storage, as well as three laboratory of micro-structurally investigation, the energetic applications of superconductivity and determination of radioisotopes concentrations. Through the important equipment of the new facilities we can enumerate: the liquefaction system with two output circuits, SEM system of scanning electronic microscopy, SQUID magnetometer, sputtering coating system, measurement system, or spectrometer with installation for radioisotopes analyses.

The results obtained till now were recognized by the national scientific research authorities for the best technological transfer of the year 2008: "The Technology of tritium extraction from heavy water used in the function of CANDU reactor, used at Nuclear Power Plant from Cernavoda". So, this isotope, tritium, which is wastage for the nuclear fission reactors, becomes fuel for the new generation of nuclear fission reactors.

At a European-scale, it has to be mentioned the fact that Romania has accelerated the rate renewable penetration, wind and photo-voltaic power plants and them integration into grid system will determine the deployment of new storage facilities. The management of the electrical system becomes more complex, once with the potential development of new nuclear power plants in Romania. In this context, the new facility named Energy Storage Technologies Laboratory (ROM-EST) was developed in addition to National Center for Hydrogen and Fuel Cell, and a new series of researches from the energy storage field, including thermal, electrochemical, or chemistry storage directions are able now to research, develop, demonstrate, or innovate them ideas. The research projects will be orientated on the development of innovative technologies for energy storage.

The laboratories will complete the portfolios of the innovative power technologies for energy storage in ICSI Rm. Valcea. The utilization of the equipment and technologies of the research of the energy storage are multidisciplinary and complex. The energy storage is essential for balancing the supply in electrical system and grid. In the future energy systems with a minimal emissions of carbon, the renewable intermittent energy, the maintenance of the electricity balance will be more complicate. Consequently, high values for the capacity of storage energy will be absolutely necessary to ensure the flexibility and stabilization of the network. The seasonal and daily fluctuation anticipations and energy storage will help the energy systems in the near future.

Figure 6.4 CRYO-HY laboratory.

The National Center for Hydrogen and Fuel Cell have the target to be a research and technological development pole in the field of hydrogen and fuel cell technologies, not just in the sense of ongoing research projects through its researchers team such as through the multitude of the services and facilities, which it is offered to other interested national and international research organizations. There are a number of testing and validation equipment extremely specific and which were purchased in order to create a national and regional facility for hydrogen research and development community.

6.6 Romanian Association for Hydrogen Energy

All peoples, not only scientists are agree with idea that a hydrogen economy, or at least one energetically system using hydrogen will be a solution for futures and will solve a series of problems such as safety, pollution, or climate changes. Hydrogen offers the greatest long-term potential for an energy system, which practically produces zero emissions, based on available

Figure 6.5 ROM-EST laboratory.

sources. Before the hydrogen reaches the assumed potential, a research-development infrastructure is needed in order to overcome an entirely unit of technical, economic, and social problems. The hydrogen technologies have a huge potential to play an important and essential role in the future energy system. Practically, any elementary source could be transformed in hydrogen, it will be possible to become a universal fuel. The hydrogen produced with renewable resources contributes to the global and implicitly the national energetically desideratum: energy security, the coordination of environment politics, or the economic competition. The utilization of hydrogen as energy vector can decrease the unstable effects of intermittent use of renewable energy source. This being said every country has a moral duty to implement both a short and medium term plans in order to turn their economies to new ones, more environmental friendly. These plans are meant to trigger research and development activities that will break any barriers on the hydrogen and fuel cell usage or other alternative and renewable energy source. All of us must understand that the first step in order to implement the hydrogen economy is the research and development activity. That is the reality for what many governments, authorities, or institutes supporting massively this field.

This is necessary to mention that the decisions to develop these types of activities is not only a scientific request but also is a social and political

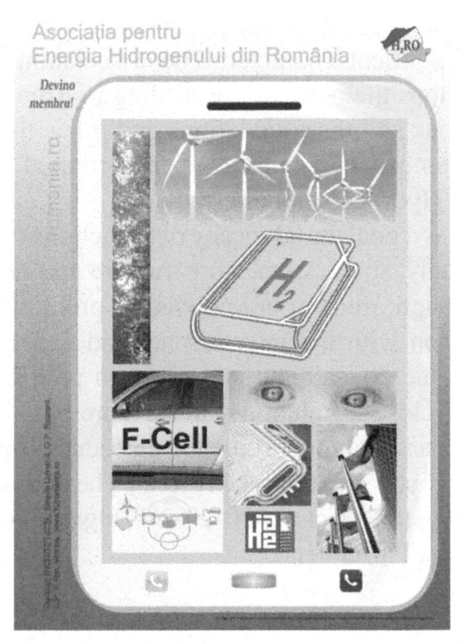

Figure 6.6 First poster of Romanian Association for Hydrogen Energy.

necessity. These research activities will bring a boost to our nation's economy value. Romania does not have yet any research, development, demonstration program for crossing over to hydrogen based on energy economy, and also no roadmap for activities that will be developed for the integration of Romania in the European effort to employment of the energy systems based on hydrogen. Those constrains determined the community to react. The scientist, and not only, concatenated the efforts and build-up Romanian Association for Hydrogen Energy, in order to disseminate knowledge to the peoples and to press the political makers.

The establishment of the association was not difficult but scattered in time. In 2006, it was inaugurated. The Romanian Alliance for Hydrogen and Fuel Cell, a structure which was supposed to become a nongovernmental organization, was composed of individual persons and legal entities. This initiative has been delayed for whatever reason both administrative and bureaucratic. In 2011, a new trial occurred, and due to some very confident persons, the approach has been successfully completed. The Registration Certificate of Romanian Association for Hydrogen Energy, as an individual person without a patrimonial purpose (NGO), was emitted by The Court of Rm. Valcea on 19th of September 2012. The association is an organization formed of individuals and legal entities, nongovernmental, with an unprofitable, autonomous, and non-political character. The association is a Romanian legal entity and it develops the activities according to Romanian legislation regarding the associations and the foundations and it has the localization in ICSI Rm. Valcea, the members being in a huge proportion of researchers from the National Center for Hydrogen and Fuel Cell.

The association has as a goal the advocacy of the actions toward the hydrogen and fuel cell economy, such as the infrastructure and renewable energy associated to these, through: transfer of knowledge, promotion of Romanian contributions, cooperation with the international and national associations, deployment support of the educational and research politics, promotion of education, all of these, in close correlation with the representation of associated members interest. The association promotes the education, research, the development, and the artistic creation or projects. The Association gives an important consideration to the "Package 20-20-20", regarding the 2009/28/CE Directive of the European Parliament.

Acknowledgments

The authors from Romania thanks National Authority for Scientifically Research, Development and Innovation from Romania http://www.research.ro. The work of them chapters was partial carried out under the contract no. 1636/2016.

References

[1] Stoft S., Dopazo C. R&D programs for hydrogen: US and EU. In: Lévêque F, Glachant J-M, Julián Barquín, Hirschhausen C, Holz F, Nuttall W, editors. Security of Energy Supply in Europe Natural Gas, Nuclear and Hydrogen, UK: Edward Elgar Publishing, Inc.; 2010, pp. 275–95.

[2] Invest East. An Overview of the Renewable Energy Market in Romania. Energy Investments & Finance; Accessed 10 April 2013. Available online at: (http://energyinvest.ro).

[3] Bozsoki I. Legal sources on renewable energy, Romania: Summary. Renewable energy policy database and support, An initiative of the European Commission; Accessed 10 April 2013. Available online at: (http://www.res-legal.eu/search-by-country/romania/summary/c/romania/s/res-e/sum/184/lpid/183).

[4] European Commission. Energy, transport and environment indicators: Eurostat; 2012.
[5] The World Bank, World Development Indicators 2012. World Bank; 2012.
[6] BP Statistical Review of World Energy. Accessed 10 April 2013. Available online at: http://www.bp.com/
[7] Transelectrica, Evolution of the Net Generating Capacity in the period 2013–2015, Accessed 1 March 2014. Available online at: www.transelectrica.ro
[8] Iordache I, Gheorghe AV, Iordache M, Towards a hydrogen economy in Romania: Statistics, technical and scientific general aspects. Int J Hydrogen Energy 2013;38:12231–40.
[9] Maisonnier G, Perrin J, Steinberger-Wilckens R; PLANET GbR. European Hydrogen Infrastucture Atlas and Industrial Excess Hydrogen Analysis, Part II Industrial surplus hydrogen, markets and production. Roads2HyCom; 2007 March. DELIVERABLE 2.1 and 2.1a. Document Number: R2H2006PU.1.
[10] Ionescu C. Study on the implementation of fuel cell in Romania (in Romnian), in Ministry of Education snd Research, Fuel Cell, vol. 1. Ed. Universul Energiei 2005, 119–133.
[11] Schervan A, Delfrate A, Roibu C. Energy production from hydrogen co-generated from OLTCHIM chlor-alkali plant by the means of fuel cells system: 2 MW installation case study. In: Sandulescu GM, editor. Proceeding of the H2_FUEL_CELLS_MILLENIUM_CONVERGENCE; 2007, Sep 21–22; Bucharest: Publishing House IPA; 2007; f31 p. 1–9.
[12] Lord A. Overview of geologic storage of natural gas with an emphasis on assessing the feasibility of storing hydrogen. Sandia National Laboratories; 2009. Raportno. SAND2009-5878. Contract No.: DE-AC04-94AL85000. Sponsored by the Department of Energy.
[13] Iordache I., Schitea D., Gheorghe A., Iordache M., Hydrogen underground storage in Romania, potential directions of development, stakeholders and general aspects, International Journal of Hydrogen Energy, 39, 11071–11081, 2014.
[14] Marban G., Valdes-Solis T., Towards the hydrogen economy?. Int J Hydrogen Energy 2007;32:1625–37.
[15] FCH JU, NOW GmbH. A portfolio of power-trains for Europe: a fact-based analysis. McKinsey & Company; 2011.

[16] Ministry of Internal Affairs (Romania). Directorate for driving licenses and vehicle registration), 2013 [2014 March 10]. Available from: http://www.drpciv.ro/

[17] Moton J., James B., Colella W. Design for manufacturing and assembly (DFMA) analysis of electrochemical hydrogen compression (EHC) systems, Proceedings of EFC2013, Fifth European Fuel Cell Technology & Applications Conference–Piero Lunghi Conference; Published by ENEA 2013; 297–8.

7

H_2 Technologies in Russia

Vladimir Fateev[1] and Sergey Grigoriev[2]

[1]National Research Center "Kurchatov Institute", Kurchatov sq., 1, Moscow, 123182, Russia
[2]National Research University "Moscow Power Engineering Institute", Krasnokazarmennaya str., 14, Moscow, 111250, Russia

7.1 Historical Milestones

The first practical application of hydrogen as a fuel took place during World War II. Hydrogen was produced by reaction of water vapor with iron for barrage balloons. An officer and an engineer B.I. Shelisch in blocked Leningrad (Saint Petersburg) offered to use the residual hydrogen from barrage balloons, which lost their volatility, as a fuel for small trucks (GAZ-AA) instead of gasoline [1]. These trucks were equipped with winch for lifting the balloons (Figure 7.1). During 1941–1942 about 200 of trucks were converted into hydrogen ones. Hydrogen from the balloon was supplied to the suction manifold of the engine through the flow tube to the working cylinders bypassing the carburetor. The dosage of hydrogen and air were provided with the throttle valve or accelerator pedal. Therefore, the control of the engine operation was similar to the control of a gasoline engine. Certainly, the efficiency was rather low, but it was the first step.

During seventies of the past century, The Central research and development automobile and engine institute NAMI (Moscow), The A.N. Podgorny Institute for Mechanical Engineering Problems NAS of Ukraine (Kharkov, Ukraine) [2], the technical university at Plant Likhachev–ZIL (Moscow State Industrial University at present time), and some others started the development of vehicles with hydrogen internal combustion engines. In 1979, Institute NAMI developed and successfully tested an experimental mini-bus RAF, which could use hydrogen and hydrogen/gasoline mixtures as a fuel (Figure 7.2) [1, 3].

Figure 7.1 Post of air defense of Leningrad front with a hydrogen "power plant" (1941) [1].

Figure 7.2 Hydrogen mini-bus RAF 22034 (1979) [3].

The further activity in hydrogen-powered vehicles with internal combustion engines was mainly concentrated in hydrogen additions to the gasoline. The main goal was a decrease of toxic wastes. In 2005, a mini-bus "Gazel" using such a mixture was demonstrated at one of the conferences of Russian Academy of Sciences (RAS) (Figure 7.3). The systems of hydrogen onboard storage (mainly metal hydrides) [6] and even catalytic systems for hydrogen onboard production were developed by Moscow Power Engineering Institute (Technical University), Institute of catalysis of RAS (Novosibirsk), Russian Federal Nuclear Center-All Russian Research Institute of Experimental Physics (Sarov), and plasma-chemical systems for gasoline conversion were developed by Kurchatov Institute of Atomic Energy (NRC "Kurchatov Institute" at present time) [4–9].

Figure 7.3 Hydrogen-gasoline mini-bus "Gazel" (2005).

But the progress in the catalytic systems of waste gases purification and absence of hydrogen infrastructure (including fueling stations) did not permit to attract a necessary attention of industry to these R&D projects.

The maximum achievement in the field of "hydrogen engines" was the creation of hydrogen aircraft engine for aircraft TU-155 and the experimental aircraft itself [10, 11]. The aircraft used liquid hydrogen as a fuel. The hydrogen engine thrust was 103 kN and duration of the flight with cryogenic fuel could reach 2 hours. The first flight was done in 1988, but the tests of such an aircraft were not long (five flights including one international flight) and the aircraft was modified for liquid natural gas fuel.

USSR space projects resulted in the development of fuel cells. Development of alkaline fuel cells was started in sixties of the past century in Russian Space Corporation "Energy" (Moscow region) together with the Ural Electrochemical Plant (Novouralsk) for soviet moon program. The Ural Electrochemical Plant (Novouralsk) together with Russian Space Corporation "Energy" developed very efficient alkaline fuel cells (i.e., "Volna" with flooded electrolyte and power 1 kW) for the space program, including fuel cell with a power 10 kW for the Soviet Space Shuttle "Buran-Energia" with efficiency 60% (matrix electrolyte, current density 0.22 A/cm^2) and the lifetime 5000–7000 h (Figure 7.5) [12–15]. The development of alkaline fuel cells was not intensive like all over the world, but such fuel cells are still

Figure 7.4 The aircraft TU-155 with hydrogen gas turbine power unit NK-88 (1989) [10].

Figure 7.5 Space-shuttle Energy-Buran (1988) and a fuel cell "Photon".

produced by the plant of electrochemical converters (The Ural Electrochemical Plant) [16].

In parallel, NPO Kvant (Moscow) were developing alkaline fuel cells for buses and even for submarines. In 1982, a prototype of such a bus RAF (produced by a plant in Riga [Latvia] with such fuel cells) was built and successfully tested [1]. The first (in the world) submarine named "Katran" with alkaline fuel cells (power 280 kW) and cryogenic hydrogen and oxygen storage was successfully tested in 1988 (Figure 7.6) [15, 17]. Later, NPO "Kvant" started to develop PEM fuel cells based on Russian polymer electrolyte membrane MF-4-SK (Nafion® type). But now NPO "Kvant" stopped the development of fuel cells.

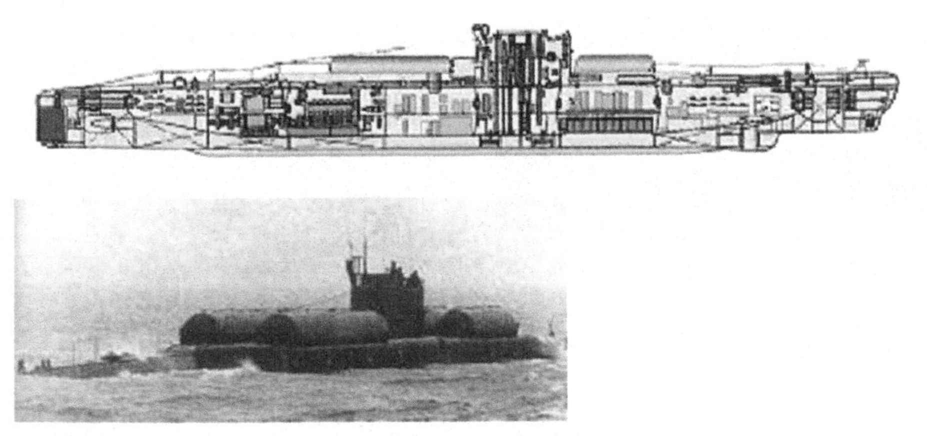

Figure 7.6 Submarine "Katran" with alkaline fuel cells and cryogenic reactant storage [17].

NRC "Kurchatov Institute" joints this R&D on PEMFC some years later. Some activity in phosphoric acid fuel cells development started at the same time but without any significant scientific and practical results. The development of solid oxide fuel cell (SOFC) started later, but in 1989 a first laboratory SOFC with a power 1 kW was built in the Institute of High Temperature Electrochemistry of Ural branch of RAS (Ekaterinburg). This SOFC was operating on natural gas at 950°C and had an efficiency of 43% and specific power *ca.* 200 mW/cm^2. That time it was the largest SOFC stack in Europe [15, 20]. Though political and economic changes in USSR stopped a lot of projects, including the polymer membrane production, these activities created a proper background for the further fuel cell development.

In 1999, the largest Russian car company AutoVAZ started developing hydrogen vehicles with alkaline fuel cells [18]. Before 2003, several experimental vehicles were built (Figure 7.7). The fuel cell was placed in a motor compartment and the hydrogen tanks in the luggage one.

As a power supply the fuel cells developed for space shuttle "Buran-Energia" were used. It is worth to stress that after a rather long storage ("on a shelf"), they demonstrated very good parameters but certainly their price was not suitable for the car industry. An electrochemical method of air purification from CO_2 was developed [19] and possibility of the use of air in alkaline fuel cell was demonstrated.

Alkaline electrolyzers were produced in USSR by a large plant "Uralkhimmash" starting from 1949. This plant was a monopolist in the production of electrolyzers in USSR and all so called "socialist countries". Such an exclusive position resulted in a very slow alkaline electrolysis technology development,

Figure 7.7 ANTEL-1 (*left*) and ANTEL-2 (*right*) produced by the car company VAZ with alkaline fuel cells [18].

and though this plant is still producing electrolyzers with productivity up to 250 m^3/h and pressure up to 10 atm [21], they cannot efficiently compete with electrolyzers of foreign companies (i.e., Hydrogenics, Ener Blue). PEM electrolyzers development was started at the end of seventies in Kurchatov Institute of Atomic Energy (National Research Center "Kurchatov Institute" at present time), and experimental electrolyzers with productivity up to 10 m^3/h were built and successfully tested [22–24]. At the same time high-temperature solid oxide electrolyzers development was started by the National Research Center "Kurchatov Institute" together with the Institute of High Temperature Electrochemistry of Ural branch of RAS. Later, this activity was mainly concentrated in the Institute of High Temperature Electrochemistry of Ural branch of RAS (Figure 7.8).

Catalytic and plasma-chemical convertors of natural gas and other organic fuels with different productivity were mainly developed by the Institute of Catalysis RAS and NRC "Kurchatov Institute" [25–28]. It is worth to stress that though plasma conversion needs some electric energy (*ca.* 0.1 kW per 1 m^3 hydrogen), it has very high productivity and is not sensitive to the fuel type and catalyst poisoning is excluded [7, 25, 26].

A very promising technology of vapor overheating by burning hydrogen for increase of turbine efficiency was developed [29–31] in Joint Institute of High Temperature RAS with assistance of Keldysh center (Moscow) [32]. This technology has a very high efficiency of hydrogen use. The increase of turbine efficiency is only several percent but the overall effect could be very high taking into account the total electric energy production by turbines. Very compact experimental steam generators were developed and successfully tested (Figure 7.9).

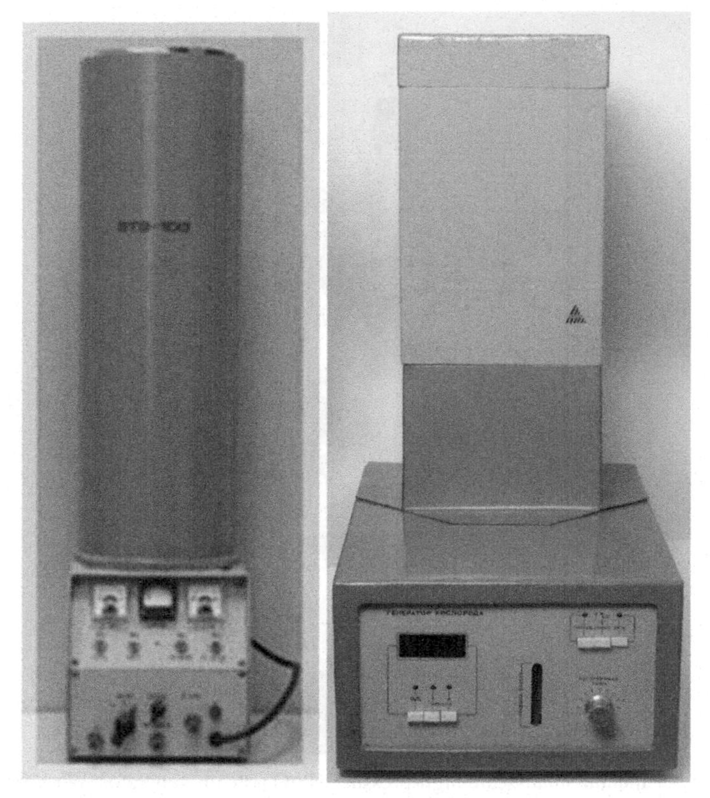

Figure 7.8 High-temperature electrolysis for 100 (*left*) and 300 (*right*) liters of hydrogen per hour developed by the Institute of High-Temperature Electrochemistry of Ural Branch of RAS [15, 20, 24].

Hydrogen safety problems were rather intensively investigated in the Joint Institute of High Temperature RAS and NRC "Kurchatov Institute" [33–36]. Both research centers had unique installation for large-scale experiments with hydrogen burning and explosion.

Metal hydrides of different types were developed and investigated in NRC "Kurchatov Institute" [6, 30], Joint Institute of High Temperature RAS, Chemical Department of Lomonosov Moscow State University, and The Institute of Problems of Chemical Physics RAS (Chernogolovka) [37–40].

It is worth to stress that during all the political and economic changes ("perestroika") due to a conversion of the industry, a number of strong research centers and institutions from the Ministry of Nuclear Energy started to develop fuel cells and other hydrogen technologies (see *i.e.*, [6, 15]). Russian-American

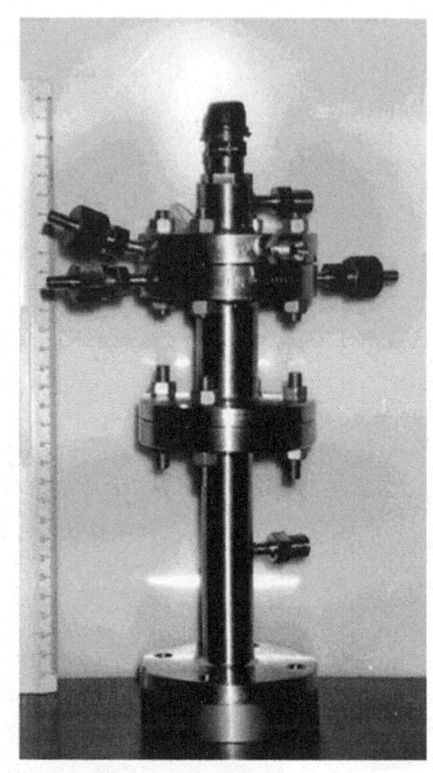

Figure 7.9 Experimental small-scale high-pressure H_2/O_2-steam generator. Working pressure, MPa: 1–4, steam temperature, κ: 600–1,000, thermal capacity, kW: 40–156, length, mm: 300, max diameter, mm: 90.

fuel cell consortium was organized [41] with their participation, and in the frame of such a collaboration, SOFC got a significant development (especially in Russian Federal Nuclear Center–Zababakhin All-Russian Scientific Research Institute of Technical Physics [Snejinsk] [42]). The main activity was concentrated on the development of SOFC power plants with power 1–5 kW with natural gas as a fuel. Such power plants were developed for the largest Russian gas company-"Gasprom" for cathodic protection of the gas pipe-lines, control systems energy supply, and so on. SOFC modules of tubular structure with the power up to 500 W were developed (Figure 7.10) with assistance of the Institute of High Temperature Electrochemistry RAS.

At the end of this "historical review," we would like to mention that in the mid of seventies of the past century in Kurchatov Institute of Atomic Energy (National Research Center "Kurchatov Institute" at present time) in

Figure 7.10 Electrochemical module of SOFC with power 500 W.

collaboration with some other research institutes, a concept of hydrogen-nuclear energy and technology was developed [43]. Hydrogen energy development was headed by the academician V. Legasov and later by academicians N. Ponomarev-Stepnoy and V. Rusanov. It gave a powerful impulse for hydrogen energy development in the USSR through the high temperature nuclear reactors, which were the basis of this concept but still are under development.

In 1996, Russian-American consortium on fuel cells was organized. This consortium was operating not too long (about 8 years) but a rather good collaboration between Russian research centers and USA National laboratories was created.

Another strong impulse for hydrogen energy development was in 2003 when one of the richest Russian companies "Norilskiy Nickel" (main precise metals producer) signed an agreement with the Russian Academy of Sciences and started financing development of the most perspective fields of hydrogen energy [44]. In 2006, this company even bought a controlling block of shares of USA company Plug Power, which was one of the leaders in the field of fuel cells that time. But after several years due to economic problems, this project was stopped though it created a background for further hydrogen energy development in a lot of Russian research centers, institutes, and universities.

At the same time (in 2003), the National association of hydrogen energy was founded. It was not a Federal organization with a target financing but it helped a consolidation of Russian research centers and universities and later concentrated its activity on codes and standards for hydrogen safety [45]. This activity appeared to be rather important as old standards could be a strong limitation for the further hydrogen energy development. For example, old standards did not permit to build any type of electrolysis plant near habitable house, so any autonomous energy supply with the use of renewables and hydrogen energy was under the question.

7.2 Present Status of R&D in Hydrogen Energy

At present time, Federal support of hydrogen energy development is reduced. Earlier (up to 2014), the hydrogen energy was one of the priorities and now it is a part of the renewable energy systems and hydrogen technologies. The main reason of this change is an unfavorable financial situation and low price for oil. But some decrease of R&D activity in hydrogen energy in USA also plays an important role. It is worth to stress that in case of USA, the development of main necessary hydrogen technologies, power plants was done and they came to the stage of hydrogen energy commercialization. Russia did not reach such a level but paid attention to the changes in R&D priorities in USA. But activity in hydrogen energy in Russia is still high and we will mention only the main directions of such an activity.

In the past few years, R&D activity was concentrated mainly on the fundamental projects and the development of new materials and technologies (not power plants) in different regions. Among the fundamental and applied research projects, nanostructured catalysts and electrocatalysts for fuel cells, electrolyzers and catalytic systems are under development mainly in NRC “Kurchatov Institute” in close collaboration with Moscow Power Engineering Institute [46–50], the Institute of Problems of Chemical Physics RAS (Chernogolovka) [51–53], Ioffe Institute RAS (Sankt Petersburg) [54, 55], the Institute of Catalysis RAS (Novosinirsk) [56–58], Frumkin Institute of Physical Chemistry (Moscow) [59–61], and the Southern Federal University (Rostov-on-Don) [62–65]. The catalysts for hydrogen production by different conversion processes (i.e., from methane, methanol) are mainly developed in Institute of Catalysis RAS [56, 57].

A significant attention is paid to nanostructured electrocatalysts on different carbon carriers, including graphene, nanofibers, nanotubes, and different other carbons like “Sibunit” [28] and the role of morphology of carbon carriers [48, 52, 53, 60]. The catalysts based on alloys, containing reduced amount of precious metals [47, 61, 62, 64], and not platinum catalysts [59] are under development. Different chemical methods and physical methods (magnetron sputtering, ion implantation) [48, 50, 66, 67] are successfully used. The latter ones permit to simplify the catalyst synthesis procedure and decrease the catalyst price.

In most of the institutes mentioned earlier, membrane-electrode assemblies and experimental PEM fuel cells are also produced. PEM and alkaline electrolyzers are developed in NRC "Kurchatov Institute" and Moscow Power Engineering Institute [68–71]. NRC "Kurchatov Institute" in collaboration with the plant "Red Star" developed PEM electrolyzers with productivity up to 10 m^3/h of hydrogen, which can operate at a pressure 130 bar (Figure 7.11) and electrolyzers for pressure up to 300 bar are under development [72]. The Institute of High Temperature RAS continues high-temperature solid oxide electrolyzers development. An experimental facility with productivity 300 l/h of oxygen (99.9%) developed and tested (Figure 7.8).

Metal hydrides, composites based on metal hydrides and pilot plants for hydrogen storage, are developed by the Institute of Problems of Chemical Physics RAS, the Institute of High Temperatures of RAS, Lomonosov Moscow State University (Chemical department) [73–77]. It is worth to stress that a significant attention is paid not only to the development of new materials but also to creation of efficient storage systems as a whole [73]. NRC "Kurchatov institute" demonstrated laboratory samples of new hydrogen storage system based on glass capillaries with hydrogen content up to 11% weight [78, 79]. In such capillaries, hydrogen could be stored at a pressure up to several hundred bars and variation of the capillaries' wall thickness gives possibility for further increase of the hydrogen weight content though it may have negative influence on the volume content.

Figure 7.11 PEM electrolysis cells (*left*) and a stack for pressure 130 bar with productivity 5 m^3 normal of hydrogen per hour.

The sodium borohydride hydrogen production systems, which could be used for small fuel cells, are developed by the State Research Institute for Chemistry and Technology of Organoelement Compounds (Moscow) and the Institute of Catalysis RAS [80–83].

Alumina systems for hydrogen production by reaction with water ("hydrogen storage/transportation system") are developed mainly in the Institute for High temperatures of RAS and some others [84–86]. At present time even a term "alumina-hydrogen" energy was offered [87] and rather large-scale demonstration power plants were built (Figure 7.12). But the development of large-scale power plants based on such technologies is still under question.

The energy consumption for alumina production is rather high and alumina (or so called "activated alumina") mainly can find application, from our point of view, in a small-scale hydrogen transportation/production system as a compact material, which does not need special storage conditions. The product of alumina oxidation and hydrolysis—bemit [γ-AlO(OH)]—is a rather attractive material from the commercial point of view, but if large-scale application of alumina for hydrogen production takes place, the price for bemit will certainly go down and commercial attraction of this compound will become significantly less.

Figure 7.12 Experimental plant (100 kW) for hydrogen production from Al [88].

Significantly more attention is paid for the development of hydrogen technologies for renewable energy systems [72, 89, 90], but practical application of such systems is not realized due to gas and oil-orientated industry.

A further development of alkaline fuel cells practically stopped like all over the world as they reached very high level but are not suitable for wide-scale common application. Some activity in this field is going on in the largest alkaline fuel cell producer in Russia-Plant of Electrochemical Convertors [16]. Earlier this plant was a division of Ural Electrochemical Plant.

PEM fuel cell development is going on in NRC “Kurchatov institute”, the Institute of Problems of Chemical Physics RAS, Krylov Research Center (Sankt Petersburg) [91], Plat “Red Star”. PEMFC experimental power plants with a power up to 10 kW were built relatively long ago [15, 92] and still the specific weight and volume parameters together with efficiency of PEM power plants are not so good as such parameters reached by the world leaders, though some engineering and technical solutions for these power plants are rather attractive. Some small innovation companies [93] are trying to organize small-scale production of PEM fuel cells for different fields of application.

SOFCs are under development in Russian Federal Nuclear Center—Zababakhin All-Russian Scientific Research Institute of Technical Physics [42], Institute of High Temperature Electrochemistry RAS, Corporation TVEL [94], Institute of Solid State Physics (Chernogolovka) [95, 96], and Institute of Electrophysics RAS (Ekaterinburg) [97, 98].

SOFC 5 kW power plant with SOFC modules 2.5 kW (Figure 7.13) for operating on natural gas and air was built in All-Russian Scientific Research Institute of Technical Physics in 2004–2006 with assistance of Institute of Electrophysics RAS and some other institutes and companies. The main direction of R&D of these teams is to exclude precise metals use in SOFC systems.

Institute of High Temperature Electrochemistry with assistance of Corporation TVEL (nuclear industry) built several demonstration power plants with power up to 1.5 kW (Figure 7.14) for Company GASPROM for cathodic protection of pipelines. Their SOFC modules were based on platinum catalysts. But up to now practical application of SOFC systems for Company GASPROM is absent and characteristics of developed SOFC power plants are not too impressive though new electrolytes and electrode materials, construction materials are successfully developed by the Institute of High Temperature Electrochemistry, Chemical Department of Moscow State University, Institute of Solid State Physics RAS, Institute of Electrophysics RAS. Development of

Figure 7.13 SOFC module 2.5 kW with tubular cells (cathode on outer side and filled in anode current collector). Specific power 300 mW/cm^2 at 0.5 V and 950°C.

Figure 7.14 SOFC-based power plants with power 1.5 kW, efficiency not less than 35%, operation using natural gas and air [99].

planar SOFC fuel cells was started in the Institute of Solid State Physics RAS not long ago.

At the end of this part, it is worth to stress that several regular conferences on hydrogen energy and fuel cells are taking part in Russia and several journals, publishing materials on these topics (see list of references), are issued. But the conferences and journals are mainly in Russian (with brief English annotations).

7.3 Hydrogen Production in Russia

The market share of hydrogen in production output of industrial gases in Russia is about 15% (Figure 7.15). This gas has almost unlimited source of raw materials [100].

About 90% of hydrogen in Russia is produced by steam conversion of natural gas (methane), and the remaining 10% by water electrolysis. Around 1,000 units of industrial-scale water electrolysers are in use at Russian companies. Taking into consideration the complexity of the processes and equipment, the most complex processes from earlier-mentioned is a steam reforming of natural gas. However, the hydrogen, produced from natural gas, is currently cheaper than electrolytic hydrogen. Therefore, the main method of hydrogen production in Russia is currently the method of catalytic steam conversion of natural gas [100].

Annual production turnout of hydrogen in Russia is about 4.5 million tons (more than 53 billion m^3) or about 8% of the worldwide global total. Hydrogen

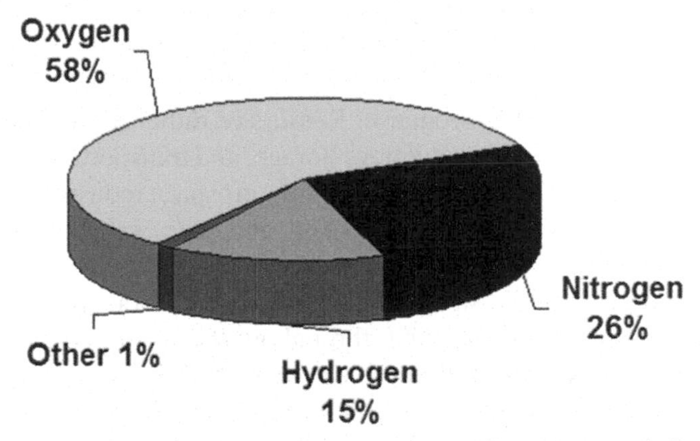

Figure 7.15 The structure of production output of industrial gases in Russia.

is not a commercial product: after the production the majority of hydrogen is used in the place of production. Unlike other industrial gases in Russia, hydrogen is most commonly produced and used at the same enterprize. The volumes of hydrogen production in Russia grow by several percent per year. Thus, according to the results of 2011–2014, hydrogen production increased by 18.5%, while in January–February 2015 Russian companies produced 114.6 million m^3 of hydrogen, which is higher than the same period of 2014 by 1.0% [101].

Almost 100% of hydrogen consumed in Russia is domestically produced. Russian foreign economic activity in the field of hydrogen is characterized by insignificant export-import operations. A foreign trade of hydrogen is almost non-existent because of transportation problems (i.e., absence of a hydrogen pipelines network). There is only a small-scale delivery of bottled gas. Such hydrogen delivery does not have a noticeable effect on the internal market. The share of imported hydrogen in 2012 was less than 1% and reaches a value of *ca.* 26,000 m^3.

It should be noted that there was no export in the same year at all. In January–February 2015, Russia has imported about 90 tons. The main hydrogen suppliers are CIS countries. A similar situation is observed in the structure of export: in January–February 2015, Russia exported 0.2 tons of hydrogen. The main buyers of hydrogen produced in Russia were CIS countries (mainly Kazakhstan) [102].

In recent years, the structure of hydrogen production in Russia has changed. The share of chemical industry decreased from 80 to 70%; thus the share of hydrogen produced by petrochemical refinery companies significantly increased. The largest increase of hydrogen production was observed in the glass industry for the period 2004–2013 when the production increased more than in three times.

Current needs for liquid hydrogen in Russia are rather limited although there is the infrastructure of its production, storage, and transportation. Liquid hydrogen is considered to be a fuel of the future. Aerospace industry of Russia is planning the creation of new carrier rockets and space systems boosters, which will use liquid hydrogen as a fuel.

It should be noted that the situation with hydrogen production in Russia differs from the situation aboard. First, throughout the world, about 40% of hydrogen is produced by gasification of coal, but the CTL-technology (Coal to Liquids) is not applied in Russia. Second, commercial hydrogen market is well developed in the world while in Russia this trend only appears. Third, the basic role in the structure of the global word-wide consumption of hydrogen takes petrochemical refining; but in Russia the main consumer of hydrogen is the industry of chemical products, especially ammonia and methanol. Fourth, it is expected that the deep processing in the refinery will be main driver of the Russian hydrogen market; while abroad, it is anticipated that increase of the hydrogen demand will be related to transport and energy sector.

There are three top regions in Russia responsible for main market share of produced hydrogen: Central Federal District (40%), Siberian Federal District (30%), and Southern Federal District (17%). For more than 5 years, the leader in terms of hydrogen production is the Kemerovo region. The total share of the Tula and Kemerovo regions is about 60% of the hydrogen produced in Russia. The largest hydrogen producer companies are JSC "Shchekinoazot", JSC "Caprolactam Kemerovo", JSC "Caustic", and JSC "Angarsk petrochemical company" [103].

Development of infrastructure network of hydrogen supply in Russia is in progress. In particular, the "Linde Gas Rus" company [104] uses two main approaches in hydrogen production and delivery: on a turnkey basis (assumed investment of the contractor) and on-site delivery (assumed investment of Linde). Today, Linde company has three plants for the production of hydrogen in Russia. Two of them are located in Ryazan and Yelabuga, working on the on-site basis for the glass industry; the third plant in located in Zelenograd (Moscow region) and produces hydrogen for the open market. The open market in Russia in the segment of small and medium-sized customers was less than 20% in the overall structure in 2013, but by 2020 it is expected to increase up to 50%. First of all it is related with the investments of international companies and the development of Russian industry (mainly in the field of flat glass and vegetable oil production). Also, the positive influence on the market will have renovation of the outdated production capacities and development of outsourcing [105].

Branch office of Hydrogenics Europe N.V. in Moscow works at the Russian market since 2002 and delivered several tens of indoor and outdoor hydrogen generators for various industries [106]. The productivity of these electrolysis systems are from 1 Nm^3 and excess 500 Nm^3.

One of the main methods of transportation of liquid hydrogen is transportation in railway tanks. The first product line of hydrogen railway tanks was developed at JSC "Uralkriomash" in Nizhny Tagil in the Soviet time for the needs of the space industry. Since then redesign of the tanks was conducted several times in order to improve the safety and key parameters of liquid hydrogen transportation. The model of the tank, which is now produced, has a number of advantages in comparison with their previous versions such as increased hydrogen transported weight, reduced losses during transportation and storage [106].

7.4 Utilization of Hydrogen in Russia

It is difficult to imagine the modern industry without the use of industrial gases in various stages of production. Today, hydrogen is one of the three most in-demand industrial gases in Russia, second only to oxygen and nitrogen. Scopes of hydrogen application in Russia are rather wide:

- food industry (hydrogenation of fats)
- chemical industry (production of ammonia, methanol, and hydrogen chloride, hydrogenation of oils)
- petrochemical industry (conversion of low-grade fuels into high-grade ones, desulphurization of fuels)
- metallurgy (creation of protective-reducing atmosphere at high temperature operations, e.g., in the production of stainless steel, cutting and welding, obtaining molybdenum and tungsten, chromium refining, obtaining hard metals based on tungsten, molybdenum, sintering of special powders, leak testing);
- energy sector (cooling of turbine generators, bubble chambers in nuclear plants)
- airspace sector (rocket fuel)
- electronics industry (creating a protective reducing atmosphere in the manufacture of semiconductors and integrated circuits)
- pharmaceutical industry (production of sorbitol)
- meteorology (to fill the pilot balloon membranes)
- glass industry (production of flat glass and quartz)
- production of pure silicon
- HF/DF chemical lasers used in industry for cutting and drilling
- in a number of other areas (Figure 7.16)

Hydrogen is one of the most important raw materials for chemical and petrochemical industry. The properties of this gas cause its application in other industrial areas such as metallurgy, food, glass, electronic, electrical sectors. Petrochemical and chemical industry uses mainly hydrogen produced by steam reforming (especially if there is a direct access to natural gas pipeline). Enterprizes of electronic, glass, food, metallurgy, and energy sector use electrolytic hydrogen because of the simplicity and reliability of water electrolyzers, the high purity of the electrolytic-grade hydrogen, possibility of

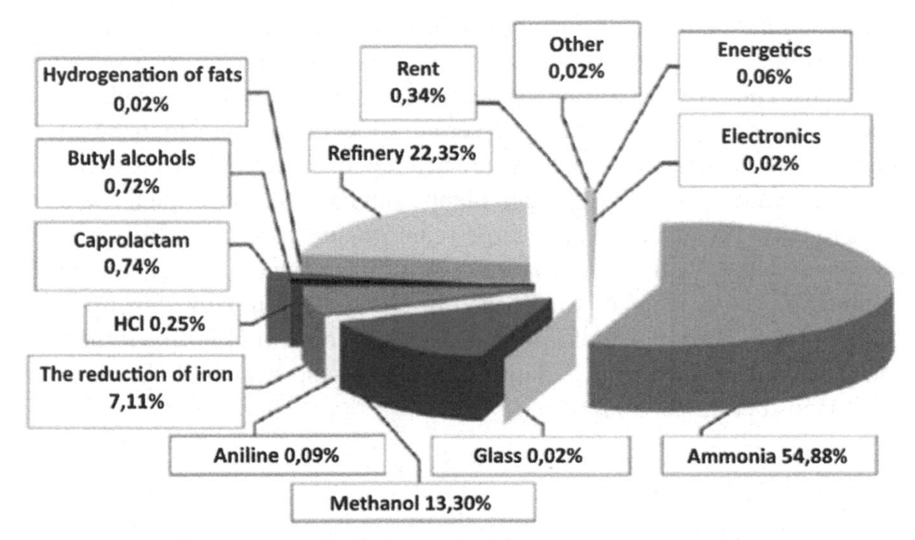

Figure 7.16 The structure of the hydrogen consumption in Russia.

generation of high-pressure gases directly in the electrolyzer, high degree of automation process and operational life.

7.4.1 Chemical Industry

The main consumer of hydrogen in Russia is the industry of chemical products, first of all, ammonia and methanol. Leaders in hydrogen consumption are enterprizes, which produce ammonia. At the moment 28 chemical plants in Russia consume *ca.* 2.5 million tons of hydrogen per year. Consumption of hydrogen in the synthesis of methanol was about 600,000 tons in 2013. Consumption of hydrogen in the remaining segments of the chemical industry (production of hydrogen chloride, hydrogenation of oils) does not exceed 90,000 tons (3% of the total consumption).

7.4.2 Oil Industry

The need for hydrogen at oil-processing plants increases from year to year. Hydrogen is used for receiving fuels from heavy high-sulphurous raw materials. A huge amount of hydrogen required for the hydro-desulfurization units, hydrocracking of distillates, hydrotreating, isomerization, production of lubricants. In addition, hydrogen is used in refineries for the activation and regeneration of reforming isomerization catalysts.

In recent years, Russia had a positive trend in the development of refining and petrochemical industries where hydrogen is used in various processes as raw materials. The existing facilities are renovated and new ones are launching. Hydrocracking and hydrotreating processes are widely implemented in the oil refining, and petrochemical companies are actively cooperating with Russian and foreign manufacturers of hydrogen plants, and increasingly implemented on-site projects. An example of a large-scale on-site project in Russia is Linde's joint venture with "Kuibyshev Azot" for the production of hydrogen and ammonia. The agreement was signed in 2013, the company start-up is expected in 2016. The volume of investments will amount to 11 billion rubles. The designed capacity of the installation is 120 thousand Nm^3/h of hydrogen and 1,340 tons per day of ammonia.

7.4.3 Metallurgy

The main application of hydrogen in the steel industry is the production of metal materials by direct reduction of iron. Now the process consumes approximately 320,000 tons of hydrogen. Significant amounts of hydrogen are consumed in the processes of rolling production (during the heat treatment of cold-rolled). Hydrogen consumption is about 15,000 tons per year. Hydrogen at the metallurgical enterprizes is also used to create a protective nitrogen-hydrogen atmosphere at high-temperature operations, such as heat treatment of pipes, stainless steel production. Hydrogen is also used in the refining of chromium, for welding and cutting.

7.4.4 Energy Sector

Due to its high thermal conductivity and diffusivity, as well as the absence of toxicity, hydrogen is used as a cooling agent in high-power turbo-generators and nuclear reactors. According to estimations, electric generation plants (including nuclear ones) consume about 4–5 thousand tons of hydrogen per year.

7.4.5 Glass Industry

In the glass industry, hydrogen is used for the plat glass production by float method, as well for quartz glass production, which is manufactured by melting of pure rock crystal, quartz, or synthetic silicon oxide in hydrogen-oxygen flame. In the glass industry, the largest consumer of electrolytic hydrogen is a glass factory in Bor, Nizhny Novgorod region. Two electrolysis installations with productivity on 500 Nm^3 H_2/h-each are used in Bor Glass Factory. This company traditionally provides glasses for cars, airplanes, and is constantly expanding its product range.

7.4.6 Food Industry

In the food industry hydrogen is used for solid fats (margarine) production by hydrogenation of oils and fats. The consumption of hydrogen in oil and fat plants is estimated at 1.5 thousand tons per year.

7.4.7 Other Consumers of Hydrogen

Among other hydrogen consumers are processing plants, factories engaged in the fabrication of nuclear fuel, the enterprize electronic and electrical industry, transportation and gas companies, pharmaceutical sector.

In 2013, in the structure of consumption of hydrogen-prevailed ammonia production (55%), followed by oil refining (22%) and obtaining methanol (13%). By 2020, the forecasted consumption growth in all these areas (up to 2.8, 2.2, and 0.8 million tons, correspondingly) [107].

Modern requirements for liquid hydrogen in Russia is extremely limited while the infrastructure of production, storage, and transportation exists. It is considered as the fuel of the future. Rocket and space industry of Russia is planning the creation of new carrier rockets and space systems boost blocks, using as fuel liquid hydrogen.

Statistics shows that during 2004–2013, the volume of hydrogen production in Russia increased in the manufacture of glass (more than 3 times). In general, the share of hydrogen production in the chemical industry decreased from 80 to 70%.

7.5 Prospects of Hydrogen Production and Utilization in Russia

A significant problem in the commercialization of hydrogen technologies in Russia is as a general unfavorable economic situation, as well the existence of its own large oil and gas reserves. At current prices of oil, investment in the development of hydrogen energy, of course, will be quite modest for some period of time. At the same time, hydrogen energy is no longer considered in Russia as one of the priority areas and was formally combined with renewable energy. Renewable energy also demonstrates slow technology growth because of oil and gas excess. However, there are some positive trends. They include the development of fuel cell-powered pilotless aircrafts [108, 109], the development of the hydrogen plant for pilot renewable energy, and backup power systems. A significant impact for the development of hydrogen technologies in Russia could be to provide the task of creation of autonomous power supply systems based on renewables with hydrogen accumulation for the Northern (including Arctic) regions [89]. In particular, development of network of tidal hydroelectric stations in the northern areas [110] requires energy storage systems, and use of hydrogen technologies for to smooth the fluctuation of energy production, apparently, has no alternative. However, all this requires more favorable economic situation.

It should be noted that the capacity of existing hydro and nuclear power stations allows to organize large-scale production of hydrogen by electrolysis of water, and here the main obstacle is the lack of the necessary internal market. Apparently, the successful development of hydrogen technology in Europe and Asia (especially in Japan) can initiate this very large and perspective market segment.

Technical regulations of hydrogen technologies in Russian Federation are engaged by Rosstandart within the framework of technical committee for standardization “Hydrogen Technology” (TC 029). By now, Russia has introduced more than 30 national standards in the field of hydrogen technologies, most of which were developed on the basis of international standards ISO [111]. In the near future, the number of implemented standards may grow twice. In 2003, National Association for Hydrogen Energy (NAHE RF) was established in Russia. The task of NAHE RF is a consolidation of various social forces for the development of a hydrogen economy, including implementation of standards [112]. It should be noted that the older standards can be a significant obstacle to the development of new hydrogen technologies, especially considering the previously laid down in the standards for hydrogen safety requirements.

This review was done at the expense of the Russian Science Foundation (project number 14-29-00111).

References

[1] Ramenskiy, A. Yu., Shelichsh, P. B., Nefedkin, S. I. Application of hydrogen as a fuel for internal combustion engines: History, present time and perspectives. International Scientific Journal for Alternative Energy and Ecology. 2006; 11(43): 63–70 (in Russian).

[2] Mischenko, A., Savitskiy, V., Baikov, V. et al. Hydrogen lift-truck Voprosi atomnoy nauki i techniki. Ser. atomno-vodorodnaja energetica i technologija. -1978, No. 3, pp. 44–45 (in Russian).

[3] http://nami.ru/

[4] http://ogaze.ru/article/vodorod-kak-prisadka-k-standartnomu-toplivu-dvs-prosto-dobav-vody

[5] http://cor.edu.27.ru/dlrstore/22b68f31-c163-10be-d8f4-824a679ab23b/b08-13_04_2003.pdf

[6] Brizitskiy, O., Terentjev, V., Khristolubov, A. et al. Development of compact devices for syn-gas on board production from hydrocarbon fuel with the purpose of fuel economy and an increase of ecological parameters of the vehicles. International Scientific Journal of Alternative Energy and Ecology. 2004; 11(19): 17–23 (in Russian).

[7] Korobtsev, S. Modern methods of hydrogen production. International chemical summit, Moscow, July 1–2, 2004 (http://www.slideserve.com/channing-massey/5917754).

[8] Vishnjakov, A. V., Jakovleva, N. V., Chastchin, V. A., Fateev, V. N. Thermodynamic and kinetic aspects of onboard hydrogen production. Khimicheskaja Technologija, No. 1, 2002. P. 3–

[9] Korobtsev, S. V., Rusanov, V. D., Kornilov, G. S., Fateev, V. N. Russian concept of ecologically clean city transport. Proceedings of International Symposium "HYPOTHESIS III", St. Petersburg, July 5–8, 1999.

[10] Sadchikov, A. "TU-155—cariogenic aviation". http://dron-sd.livejournal.com/26358.html

[11] FLIGHT INTERNATIONAL, 18 February 1989. https://www.flightglobal.com/FlightPDFArchive/1989/1989%20-%200410.PDF

[12] Matrenin, V. I., Ovchinnikov, A. T., Pospelov, B. S., Sokolov, B. A., Stikhin, A. S. From power system of buran orbiter to power system of space vehicles and stations. Kosmicheskaja tehnika i tehnologii. 2013; 3: 57–65.

[13] Grigorov, E. I., Korovin, N. V., Khudyakov, S. A. "Hydrogen-oxygen fuel cell plant for ship "Buran." Abstracts of 42nd Annual Meeting of International Society of Electrochemistry, Montreux, Switzerland, 1991, 1–18.

[14] Khudjakov, S. Development of power plants based on alkaline fuel cells for moon orbital space ship and multiuse spaceship "Buran." International Scientific Journal for Alternative Energy and Ecology. 2002; 6: 37–62 (in Russian).

[15] Kozlov S. I., Fateev V. N. Hydrogen Energy: Modern Status, Problems and Perspectives. 2009. Gasprom VNIIGAS, 520, Moscow.
[16] http://www.novozep.ru
[17] Nikiforov, B., Sokolov, V., Sokolov, B., Khudjakov, S. New power sources for submarines. International Scientific Journal for Alternative Energy and Ecology. 2002; 2: 15–19 (in Russian).
[18] Kondratiev, D., Matrenin, V., Pospelov, B., Stikhin, A. Power sources of ZEP OAO "UEHK." Jurnal Avtomobilnikh Injenerov. 2012; 74(3): PP. 36–39 (in Russian).
[19] Bolshakov, K., Kondratiev, D., Matrenin, V., Ovchinnikov, A., Pospelov, B., Potanin, A., Stokhin, A. Electrochemical method of air purification from CO_2. Electrochemical Eenergetics. 2009; 9(3): 147–151.
[20] Somov, S., Demin, A., Kuzin, B., Lipilin, A., Perfiliev, M. "High temperature electrochemical generator." Russian patent No. 2027258, 20.01.1995.
[21] http://ekb.ru/
[22] Grigoriev, S. A., Kalinnikov, A. A., Millet, P., Porembsky, V. I., Fateev, V. N. Mathematical modeling of high-pressure PEM water electrolysis. Journal of Applied Electrochemistry. 2010; 40(5): 921–932.
[23] Grigoriev, S. A., Porembsky, V. I., Fateev, V. N. Pure hydrogen production by PEM electrolysis for hydrogen energy. International Journal of Hydrogen Energy. 2006; 31(2): 171–175.
[24] http://ihte.uran.ru
[25] Rusanov, V. D., Babaritskiy, A. I., Baranov, E. I., Demkin, S. A., Jivotov, V. K., Potapkin, B. V. Plasma catalytic effect for dissociation of methane for hydrogen and carbon. Doklady Akademii Nauk. 1997; 354(2): 1–3 (in Russian).
[26] Potekhin, S. V., Potapkin, B. V., Deminskiy, M. F. Plasma catalytic effect at methane decomposition. High Energy Chemistry. 1997; 33(1): 59–66.
[27] Makarshin, L. I., Parmon, V. N. Micro-channel catalytic systems for hydrogen energy. Russian Chemical Journal. 2006; 1(6): 52–58 (in Russian).
[28] http://catalysis.ru/
[29] Malyshenko, S. P. LIHT RAS research and development in the field of hydrogen energy technologies. International Scientific Journal for Alternative Energy and Ecology. 2011; 95(3):10–34 (in Russian).

[30] Pekhota, F. N., Rusanov, V. D., Malyshenko, S. P. Russian Federal Hydrogen Energy Program. International Journal of Hydrogen Energy. 1998; 23(10): 967–970.
[31] Malyshenko, S. P., Gryaznov, A. N., Filatov, N. I. High-pressure H_2/O_2-steam generators and their possible applications. International Journal of Hydrogen Energy. 2004; 29(6): 589–596.
[32] http://kerc.msk.ru/
[33] Petukhov, V. A., Naboko, I. M., Fortov, V. E. Explosion hazard of hydrogen-air mixtures in the large volumes. International Journal of Hydrogen Energy. 2009; 34(14): 5924–5931.
[34] Golub, V. V., Baklanov, D. I., Bazhenova, T. V., Golovastov, S. V., Ivanov, M. F., Làskin, I. N., Semin, N. V., Volodin, V. V. Experimental and numerical investigation of hydrogen gas auto-ignition. International Journal of Hydrogen Energy. 34(14): 5946–5953.
[35] Grigoriev, S. A., Millet, P., Korobtsev, S. V., Porembskiy, V. I., Pepic, M., Etievant, C., Puyenchet, C., Fateev, V. N. Hydrogen safety aspects related to high-pressure polymer electrolyte membrane water electrolysis. International Journal of Hydrogen Energy. 34(14): 5986–5991.
[36] Korobtsev, S. V., Fateev, V. N., Samsonov, R. O., Kozlov, S. I. Safety of hydrogen energy. Transport on Alternative Fuel. 2008; 5: 68–72 (in Russian).
[37] Artemov, V. I., Lazarev, D. O., Yan'kov, G. G., Borzenko, V. I., Dounikov, O. O., Malyshenko, S. P. Mathematical modeling of Heat and Mass Transfer in Metal Hydrogen accumulating and purification system. Heat Transfer Research. 2004; 35(1): 140–148.
[38] Verbetsky, V. N., Malyshenko, S. P., Mitrokhin, S. V., Solovey, V. V., Shmal'ko, Yu. F. Metal hydrides: properties and practical applications. Review of the works in CIS-countries. International Journal of Hydrogen Energy. 1998; 23(12): 1165–1177.
[39] Tarasov, B. P., Moravsky, A. P., Goldshleger, N. F. Hydrogen-containing carbon nanostructures: synthesis and properties. Russian Chemical Reviews. 2001; 70(2): 131–146.
[40] Tarasov, B. P., Fokin, V. N., Moravsky, A. P., Shul'ga, Yu. M., Yartys, V. A. Hydrogenation of fullerenes C60 and C70 in the presence of hydride-forming metals and intermetallic compounds. Journal of Alloys and Compounds. 1997; 253–254: 25–28.
[41] http://www.sandia.gov/media/fuelcell.htm
[42] http://www.vniitf.ru/tverdooksidnye-toplivnye-elementy

[43] Ponomarev-Stepnoy, N. N., Stoljarevskiy, A. Ya. Nuclear-hydrogen energy—ways of development. Energy. 2004; 1: 3–9.

[44] http://www.h2fc-fair.com/hm08/images/exhibitors/nic-nep-pm-e.pdf

[45] http://www.h2org.ru

[46] Grigoriev, S. A., Fateev, V. N., Lutikova, E. K., Grigoriev, A. S., Bessarabov, D. G., Xing Wei, Junjie Ge. CNF-supported platinum electrocatalysts synthesized using plasma-assisted sputtering in pulse conditions for the application in a high-temperature PEM fuel cell. International Journal of Electrochemical Science. 2016; 11: 2085–2096.

[47] Grigoriev, S. A., Millet, P., Fateev, V. N. Evaluation of carbon-supported Pt and Pd nanoparticles for the hydrogen evolution reaction in PEM water electrolysers. Journal of Power Sources. 2008; 177(2): 281–285.

[48] Grigoriev, S. A., Fedotov, A. A., Martemianov, S. A., Fateev, V. N. Synthesis of Nanostructural Electrocatalytic Materials on Various Carbon Substrates by Ion Plasma Sputtering of Platinum Metals. Russian Journal of Electrochemistry. 2014; 50(7): 638–646.

[49] Pushkarev, A. S., Pushkareva, I. V., Grigoriev, S. A., Kalinichenko, V. N., Presniakov, M. Yu., Fateev, V. N. Electrocatalytic layers modified by reduced graphene oxide for PEM fuel cells. International Journal of Hydrogen Energy. 2015; 40(42): 14492–14497.

[50] Fedotov, A. A., Grigoriev, S. A., Lyutikova, E. K., Millet, P., Fateev, V. N. Characterization of carbon-supported platinum nanoparticles synthesized using magnetron sputtering for application in PEM electrochemical systems. International Journal of Hydrogen Energy. 2008; 38(1): 426–430.

[51] Gerasimova, E. V., Safronova, E. Yu., Volodin, A. A., Ukshe, A. E., Dobrovolsky, Yu. A., Yaroslavtsev, A. B. Electrocatalytic properties of the nanostructured electrodes and membranes in hydrogen-air fuel cells. Catalysis Today. 2012; 193: 81–86.

[52] Jaroslavtsev, A. B., Dobrovolskiy. Ju. A., Shaglaev, N. S., Frolov, L. A., Gerasimova, E. V., Sanginov, E. V. Nanostructured materials for low temperature fuel cells. Russian Chemical Review. 2012; 81(3): 191–220.

[53] Gerasimova, E. V., Bukun, N. G., Dobrovolsky, Yu. A. Electrocatalytic properties of the catalysts based on carbon nanofibers with various platinum contents. Russian Chemical Bulletin. 2011; 60(6): 1045–1050.

[54] Zabrodskii, A. G., Glebova, N. V., Nechitailov, A. A., Terukova, E. E., Terukov, E. I., Tomasov, A. A., Zelenina, N. K. Membrane-electrode

assemblies with high specific power based on functionalized carbon nanotubes. Technical Physics Letters.2010; 36(12): 1112–1114.

[55] Nechitailov, A. A., Glebova, N. V., About mechanism of the influence of oxygen-modified carbon nanotubes on the kinetics of oxygen electroreduction on platinum Russian Journal of Electrochcmistry, Vol. 50, 2014, pp. 835–840.

[56] http://catalysis.ru/

[57] Kirillov, V. A., Fedorova, Z. A., Danilova, M. M., Zaikovskii, V. I., Kuzin, N. A., Kuzmin, V. A., Krieger, T. A., Mescheryakov, V. M. Porous nickel-based catalyst for partial oxidation of methane to synthesis gas. Applied Catalysis A: General. 2011; 401(1–2): 170–175.

[58] Mishakov, I. V., Bujanov, R. A., Zaikovskiy, V. I., Streltsov, I. A., Vedjagin, A. A. Catalytic production of carbon nanosized structures of feathery morphology by mechanism of carbide cycle. Kinetics and Catalysis. 2008; 49(6): 916–921.

[59] Tarasevich, M. R., Bogdanovskaya, V. A., Kuznetsova, L. N., Modestov, A. D., Efremov, B. N., Chalykh, A. E., Chirkov, Yu. G., Kapustina, N. A., Ehrenburg, M. R. Development of platinum-free catalyst and catalyst with low platinum content for cathodic oxygen reduction in acidic electrolytes. Journal of Applied Electrochemistry. 2007; 37: 1503–1513.

[60] Bogdanovskaya, V. A., Tarasevich, M. R. Mechanism of corrosion of nanosized multicomponent cathode catalysts and formation of core-shell structures. International Scientific Journal for Alternative Energy and Ecology. 2009; 12: 24–56.

[61] Bogdanovskaya, V. A., Tarasevich, M. R., Lozovaja, O. V. Kinetics and mechanism of oxygen electroreduction on PtCoCr/C-catalyst with platinum content 20–40% mass. Russian Journal of Electrochemistry. 2011; 47(7): 902–917.

[62] Leontyev, I. N., Guterman, V. E., Pakhomova, E. B., Timoshenko, P. E., Guterman, A. V., Zakharchenko, I. N., Petin, G. P., Dkhil, B. XRD and electrochemical investigation of particle size effects in platinum–cobalt cathode electrocatalysts for oxygen reduction. Journal of Alloys and Compounds. 2010; 500: 241–246.

[63] Leontyev, I. N., Chernyshov, D. Yu., Guterman, V. E., Pakhomova, E. B., Guterman, A. V. Particle size effect of carbon supported Pt-Co alloy electrocatalysts prepared by the borohydride method: XRD characterization. Applied Catalysis A: General. 2009; 357: 1–4.

[64] Guterman, V. E., Lastovina, T. A., Belenov, S. V., Tabachkova, N. Yu., Vlasenko, V. G., Khodos, I. I., Balakshina, E. N. PtM/C (M=Ni, Cu, or Ag) electrocatalysts: effects of alloying components on morphology and electrochemically active surface areas. Journal of Solid State Electrochemistry. 2014; 18: 1307–1317.

[65] Leontyev, I. N., Belenov, S. V., Guterman, V. E., Haghi-Ashtiani, P., Shaganov, A., Dkhil, B. Catalytic Activity of Carbon Supported Pt/C Nano-Electrocatalysts. Why Reducing the Size of Pt Nanoparticles is not Always Beneficient. The Journal of Physical Chemistry. 2011; 115: 5429–5434.

[66] Alexeeva, O. K., Fateev, V. N. Application of the magnetron sputtering for nanostructured electrocatalysts synthesis. International Journal of Hydrogen Energy. 2016; 41(5): 3373–3386.

[67] Fateev, V., Alekseeva, O., Lutikova, E., Porembskiy, V., Nikitin, S., Mikhalev, A. New physical technologies for catalyst synthesis and anticorrosion protection. International Journal of Hydrogen Energy. 2016; 41(25): 10515–10521.

[68] Kuleshov, V. N., Kuleshov, N. V., Grigoriev, S. A., Udris, Y. Y., Millet, P., Grigoriev, A. S. Development and characterization of new coating for application in alkaline water electrolysis. International Journal of Hydrogen Energy. 2016; 41(1): 36–45.

[69] Kuleshov, N. V., Kuleshov, V. N., Dovbysh, S. A., Udris, E. Ya., Grigor'ev, S. A., Slavnov, Yu. A., Korneeva, L. A. Polymeric composite diaphragms for water electrolysis with alkaline electrolyte. Russian Journal of Applied Chemistry. 2016; 89(4): 618–621.

[70] Grigoriev, S. A., Kalinnikov, A. A., Millet, P., Porembsky, V. I., Fateev, V. N. Mathematical modeling of high pressure PEM water electrolysis. Journal of Applied Electrochemistry. 2010; 40(5): 921–932.

[71] Fateev, V. N., Lutikova, E. K., Pimenov, V. V., Akelkina, S. V., Salnikov, S. E., Xing, W. About possibility of a PEM electrolyzers start at negative temperatures. International Scientific Journal for Alternative Energy and Ecology. 2015; 12(176): 28–39.

[72] Fateev, V., Blach, P., Grigoriev, S., Kalinnikov, A., Porembskiy, V. High pressure PEM electrolyzers and their application for renewable energy systems European Hydrogen Energy Conference 2014. Proceedings, Seville, Spain 12–14th of March. (EHEC 2014 abstracts) pp. 129–130.

[73] Blinov, D. V., Borzenko, V. I., Dunikov, D. O., Romanov, I. A. Experimental investigations and a simple balance model of a metal

hydride reactor. International Journal of Hydrogen Energy. 2014; 39(33): 19361–19368.

[74] Tarasov, B. P., Fokin, V. N., Fokina, E. E., Yartys, V. A. Synthesis of hydrides by interaction of intermetallic compounds with ammonia. Journal of Alloys and Compounds. 2015; 645: S261–S266.

[75] Satya Sekhar, B., Lototskyy, M., Kolesnikov, A., Moropeng, M. L., Tarasov, B. P. Performance analysis of cylindrical metal hydride beds with various heat exchange options. Journal of Alloys and Compounds. 2015; 645: S89–S95.

[76] Anikina, E. Yu., Verbetsky, V. N., Savchenko, A. G., Menushenkov, V. P., Shchetinin, I. V. Calorimetric study of hydrogen interaction with Sm_2Fe_{17} Journal of Alloys and Compounds, 2015; 645(S1): S257–S260.

[77] Verbetsky, V. N., Lushnikov, S. A., Movlaev, E. A. Interaction of vanadium alloys with hydrogen at high pressures. Inorganic Materials. 2015; 51(8): 779–782.

[78] Zhevago, N. K., Chabak, A. F., Denisov, E. I., Glebov, V. I., Korobtsev, S. V. Storage of cryo-compressed hydrogen in flexible glass capillaries. International Journal of Hydrogen Energy. 2013; 38: 6694–6703.

[79] Zhevago, N. K., Glebov, V. I., Denisov, E. I., Korobtsev, S. V., Chabak, A. F. Micro-capillary vessels for hydrogen storage. International Scientific Journal for Alternative Energy and Ecology. 2012; 09(113): 106–116.

[80] http://eos.su/

[81] Varganov, B. P., Storojenko, P. A. Russian Patent "Hydrogen generator" RU No 2385288. 2010.

[82] Netskina, O. V., Ozerova, A. M., Komova, O. V., Odegova, G. V., Simagina, V. I. Hydrogen storage systems based on solid-state $NaBH_4/Co_xB$ composite: influence of catalyst properties on hydrogen generation rate. Catalysis Today. 2015; 245(1): 86–92.

[83] O. V. Netskina, R. V. Fursenko, O. V. Komova, E. S. Odintsov, V. I. Simagina $NaBH_4$ generator integrated with energy conversion device based on hydrogen combustion Journal of Power Sources, Vol. 273, 1 January 2015, P. 278–281.

[84] Larichev, M. N., Laricheva, O. O., Shaitura, N. S., Shkolnikov, E. I. Possibilities of practical usage of dispersed aluminium oxidation by liquid water. Thermal Engineering. 2012; 59(13): 1000–1009. © Pleiades Publishing, Inc., 2012. ISSN 0040 6015.

[85] Ilyukhina A. V., Ilyukhin A. S., Shkolnikov E. I. Hydrogen generation from water by means of activated aluminum. International Journal of Hydrogen Energy. 2012; 37(21): 16382–16387.

[86] Vlaskin, M. S., Shkolnikov, E. I., Zhuk A. Z. et al. Computational and experimental investigation on thermodynamics of the reactor of aluminum oxidation in saturated wet steam. International Journal of Hydrogen Energy. 2010; 35: 1888–1894.

[87] Sheindlin, A. E., Juk, A. Z., Shkolnikov, E. I. et al. Alumina-hydrogen energy Moscow, OIVTRAN. 2007, 278 p. (in Russian).

[88] http://www.jiht.ru/science/science_council/presentations/shkolnikov/report_091214.pdf

[89] Grigor'ev, S. A., Grigor'ev, A. S., Kuleshov, N. V., Fateev, V. N., Kuleshov, V. N. Combined Heat and Power (Cogeneration) Plant Based on Renewable Energy Sources and electrochemical Hydrogen Systems. Thermal Engineering. 2015; 62(2): 81–87.

[90] Kuleshov, N. V., Grigoriev, S. A., Kuleshov, V. N., Terentyiev, A. A., Fateev, V. N. Low temperature water electrolyzers for autonomous power plants with hydrogen energy accumulation. International Scientific Journal for Alternative Energy and Ecology. 2013; 6(127): 23–27.

[91] http://krylov-center.ru/

[92] Fateev, V. N., Morozov, Yu. V., Porembsky, V. I., Grigoriev, S. A. Electrochemical systems with proton exchange membranes, as a basis of a hydrogen infrastructure. h2storage.net›docs/pdf/28/1230-1250_fateev.pdf

[93] www.atenergy.pro

[94] http://tvel.ru

[95] http://www.issp.ac.ru/main/

[96] Kolotygin, V. A., Tsipis, E. V., Lu, M. F., Pivak, Y. V., Yarmolenko, S. N., Bredikhin, S. I., Kharton, V. V. Functional properties of SOFC anode materials based on $LaCrO_3$, La(Ti, Mn)O_3 and Sr(Nb, Mn)O_3 perovskites: a comparative analysis. Solid State Ionics. 2013; 251: 28–33.

[97] http://www.iep.uran.ru/

[98] Spirin, A. V., Nikonov, A. V., Lipilin, A. S., Paranin, S. N., Ivanov, V. V., Khrustov, V. R., Valentsev, A. V., Krutikov, V. I. Electrochemical cell with solid oxide electrolyte and oxygen pump thereof. Russian Journal of Electrochemistry. 2011; 47(5): 569–578.

[99] http://www.ihte.uran.ru/?page_id=155

[100] Stukalov V. A., Subbotin S. A., Shhepetina T. D. “Vodorodnaja jenergetika i tehnogennyj vodorodnyj cikl kak osnova konsolidirovannogo razvitija toplivodobyvajushhih otraslej i atomnoj jenergetiki”, RNC Kurchatovskij institut.

[101] http://www.sostav.ru/blogs/32702/17219

[102] http://www.fedstat.ru

[103] http://www.drgroup.ru/438-analiz-rinka-vodoroda-i-vodorodnoi-energetiki-v-rossii.html

[104] http://www.linde-gas.com/en/products_and_supply/gases_fuel/hydro gen.html

[105] http://neftegaz.ru/analisis/view/8190

[106] http://hydrogenics.ru

[107] http://tgko.ru/news/post_reliz_konferencii_vodorod_2014

[108] http://www.ciam.ru/?NewsId=1912&lang=RUS

[109] Baranov, I., Fateev, V., Porembskii, V., Kalinnikov, A. Aviation power plant on hydrogen-air solid polymer fuel cell. Alternative Fuel Transport. 2015; 3(45): 36–44.

[110] Usachev I. N., Shpoljanskij Yu. B., Istorik B. L. et al. “Prilivnye jelektrostancii (PJeS)—istochnik jenergii, zapasaemyj v vodorode 2-j mezhdunarodnyj forum Vodorodnye tehnologii dlja razvivajushhegosja mira. Tezisy dokladov. 2008.

[111] Ramenskiy, A. Yu., Grigoriev, S. A. Tehnologii toplivnyh jelementov: Voprosy tehnicheskogo regulirovanija [Fuel cell technologies: Technical regulation issues]. International Scientific Journal for Alternative Energy and Ecology (ISJAEE). 2016; 19–20(207–208): 107–129 (in Russian).

[112] http://h2org.ru

8

Hydrogen (H_2) Technologies in the Republic of South Africa

Dmitri Bessarabov[1] and Bruno G. Pollet[2]

[1]DST HySA Infrastructure Center (Hydrogen South Africa), North-West University, Faculty of Engineering, Private Bag X6001, Potchefstroom, 2520, South Africa
[2]Power and Water (KP2M Ltd), Swansea, SA6 8QR, Wales, UK

Abstract

Global demand for renewable hydrogen, power-to-gas, and fuel cell technologies are increasing rapidly as many countries, particularly Germany, Japan, USA, China, and South Korea, are focusing their attention to clean energy technologies, which will improve their *triple bottom-line* (financial, environmental, and social). The South African Government, through *HySA* (Hydrogen South Africa) program, as well as through business-driven approach of PGM (Platinum Group Metal) mining companies, such as *Anglo American Platinum*, *Impala Platinum,* and *Lonmin*, are aggressively pursuing research, development, and market creation for hydrogen, fuel cell, and water electrolysis technologies. These activities will result in the beneficiation of minerals in South Africa (SA), in turns increasing employment for high-quality jobs, development of human capital, and improved quality of life. In SA, there is a significant demand for energy in rural areas. This includes critical infrastructure sites such as schools and hospitals, as well as households. Extension of the electrical grid to most of these areas is currently not economically viable and feasible. This represents a great opportunity to fulfil unmet needs for on-site power generation using clean energy technology approaches. Other market opportunities for HFCT (Hydrogen and Fuel Cell Technologies) in SA include the mining and the telecom sectors (especially in the whole of the Sub-Saharan area). For these reasons, the Department of Science and Technology (DST) of SA developed the National Hydrogen and Fuel Cell Technologies (HFCT) Research, Development, and Innovation (RDI) Strategy. The National Strategy was branded Hydrogen South Africa (*HySA*). The overall goal of HySA is to develop and guide innovation along the value chain of HFCT in SA. Allied to this, one of the objectives is to support and develop a range of high-level skills, generally in accordance with the required Human Capital Development (HCD) strategy.

This chapter reports on the general activities of *HySA* in SA as well as focuses on some key hydrogen and fuel cell technologies that are under development at *HySA*. A brief overview of the general energy profile of South Africa will be highlighted and other than *HySA* players in the hydrogen space will be briefly discussed.

Keywords: Republic of South Africa, Energy, Hydrogen, Fuel Cells, *HySA*.

8.1 Introduction

Africa is home to six of the ten fastest growing economies, driven largely by foreign investment (inc. AfDB—African Development Bank) reaching ∼US$150 billion in 2015. The Republic of South Africa (RSA) is the only one out of 54 countries making up the African continent. Now with over 1.1 billion people, the continent is being seen by China, Europe, and increasingly the United States, as the next major trading partner. It is thought that 65% of the population have no access to electricity. This is very unevenly distributed across the continent, with North African countries having over 90% access to electricity, but countries in Sub-Saharan Africa (SSA) this drops to 50% or under. According to the 2009 International Renewable Energy Agency (IRENA) [1], just five countries in Africa dominate the current power market. These are South Africa at 21% of primary energy use, Nigeria at 16%, Egypt at 11%, Algeria at 6%, and Ethiopia at 5%. The rest of the African continent represents only 41% of the total primary energy use. In terms of energy and Africa, there are a few countries without a policy document looking at energy but in reality fewer of them are being acted upon. Some relevant examples [2] are as follows:

- The Nigerian government has put a legislative framework in place under the Renewable Energy Masterplan. Within this off-grid and distributed solar are actively being encouraged. According to some of the frameworks, there is a solar PV target of 500 MW by 2025.

- The Ghanaian government has a 10% renewable energy target by 2020.
- Zambia has a rural electrification scheme, with a focus on deploying renewable energy. Within this the government is offering a range of fiscal incentives.

With an annual energy consumption of ~1,100 TWh RSA, a *BRICS* middle-income developing country faces many issues to economic growth such as energy challenges and aged, underinvested, and inadequate power infrastructures. Although SA is often seen as the "Powerhouse of Africa," it is facing major energy challenges such as unplanned and planned power outages, energy shortages, power blackouts, high energy tariffs, and energy poverty in low income households. In views of kick-starting the economic growth and supplying "energy for all," the SA government has developed a "5-Point Energy Plan" comprising simultaneous actions in key strategic areas, consisting of: (i) maintaining the country's state-owned electricity company *Eskom*; (ii) introducing new generation capacity through coal; (iii) partnering with the private sector into co-generation contracts; (iv) introducing gas-to-power technologies; and, (v) accelerating the demand side management [3]. The SA has also rolled out several energy and energy efficiency programmes and initiatives with a strong emphasis on off-grid Renewable Energy solutions, Hydrogen Energy, and oil and gas (including shale) exploration opportunities [1]. The country's aspiration is to build-up to an "industrial revolution" incorporating the development of a "Green Economy," which could significantly boost the nation's manufacturing and mining base (through minerals and Platinum Group Metals (PGM) beneficiation) [3].

8.2 Current Energy Landscape in RSA

Africa is facing a major challenge: 1.1 billion people live on the continent (14% of the world population) with 65% of the population having no access to electricity. Africa is still home to the world's largest concentration of an impoverished population, and the considerable gap between "rich" and "poor" continues to widen. By 2050, Africa's population is estimated to reach 1.9 billion, and consequently, energy supply and access will continue to be major issues. In order to meet the projected growth in electricity demand, Africa needs to add ~250 GW of new capacity between now and 2030, totalling to an annual capital cost of ~US$20 billion by 2030 (see the following Figure 8.1) [3].

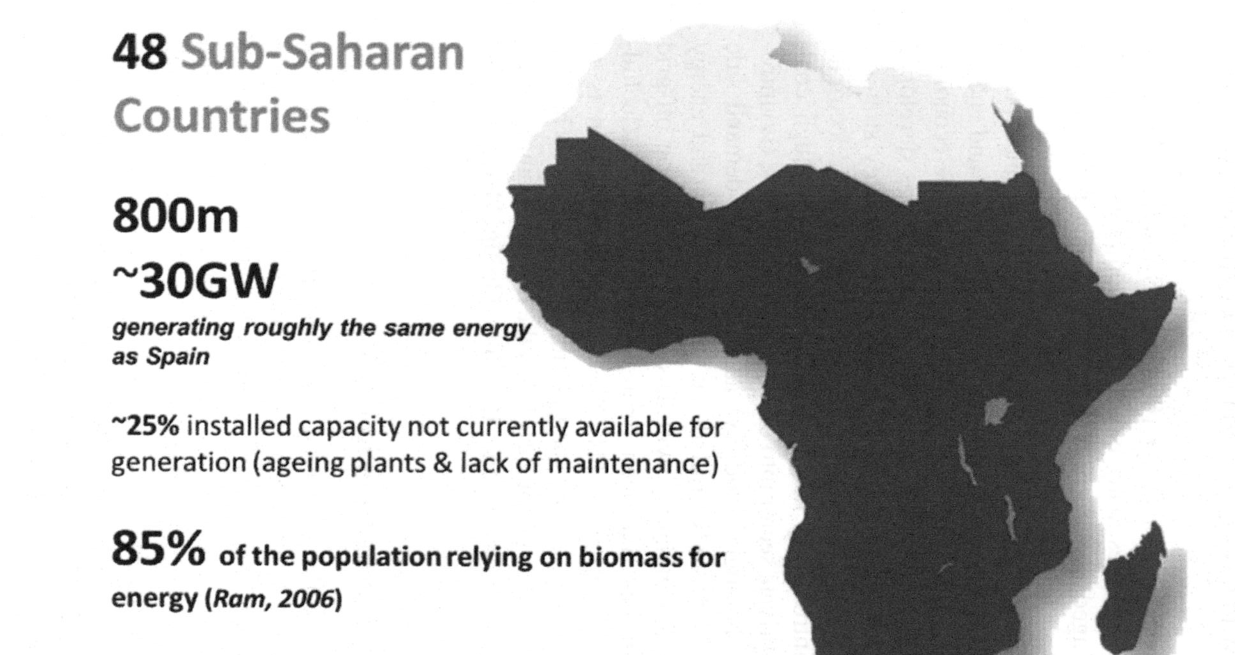

Figure 8.1 Map of Africa showing areas of energy installation and generation [4].

Source: © Bruno G. Pollet.

Although SA's abundance of coal, present lack of liquid fuel reserves, and the historic isolation of the country have created a situation where coal is the primary source of energy, SA's coal production has grown steadily by 2% per year since the early 1990s, and around three quarters of its viable coal reserves remain underground: 30.1 of 38.6 GT total (including historic production). SA's proven reserves amount to ~8% of total global coal reserves. A naive assumption is that with 260 MT annual production in 2012, SA coal reserves will last for 115 years. This neglects both the increasing rate of production and improvements in technology that allow more coal to be found and extracted at a given price.

To date natural gas in Africa has been a very minor fuel. This is likely to change over the next decade as gas has the potential to account for more than 40% of the electricity generated in Sub-Saharan Africa (SSA) from 2020 onwards, and by 2040, gas-fired capacity could be responsible for more than 700 terawatt-hours in the region. In SA, this lack of usage and interest is clearly known although an increasing focus is being given to potential reserves of shale gas in the Karoo Basin, and the SA government has now identified shale gas as a "potential game changer." SA's lower Karoo is estimated to contain 1,559 trillion cubic feet (tcf) of potential shale gas reserves of which 390 tcf is classed as (currently) technically recoverable. Putting this into context, 390 tcf has the equivalent energy content of around 13.2 GT of coal, adding around a third to SA's total reserves of fossil energy. This puts SA's potential shale reserves in the top ten in the world [3].

Around 5–6.5% of SA's electricity is provided via *Eskom*'s 1,800 MW Koeberg Nuclear Power Station's two reactors (French-built) in the Western Cape Province. The Nuclear Energy Policy (NEP) set by the state-owned Nuclear Energy Corporation of South Africa (NECSA) aims to (i) increase the role of nuclear energy as part of the process of diversifying SA's primary energy sources to ensure energy security, (ii) reduce the country's over-reliance on coal, and (iii) become globally competitive in the long-term vision for SA in the use of their innovative technology for the design, manufacture, and deployment of state-of-the-art nuclear energy systems and power reactors, and nuclear fuel-cycle systems. As part of the Energy plan, in 2014, it was announced that SA will build six new nuclear power plants by 2030, providing 99.6 GW of power at a cost estimated between US$36 billion to US$90 billion [3].

In SA, there has been an increase in the installation of solar water heaters and solar PV panels in commercial buildings and private dwellings. Southern Africa has one of the highest numbers of sunny days in the world (more than 2,500 hours of sunshine/year) and significantly higher levels of radiation (average solar radiation levels ranging between 4.5 and 6.5 kWh/m^2 per day). Solar technologies, particularly Concentrating Solar Power (CSP), are rapidly approaching grid parity in SA due both to the abundance of resource and the recent rises in electricity prices. CSP is of particular interest as it can be combined with the thermal storage to provide baseload, dispatchable power. CSP is the only sustainable and dispatchable technology that could supply SA's electricity demand, and that a balanced mix of PV, wind, and CSP can provide the energy supply needed in SA [3].

The wind resource in SA is also particularly strong, especially along the southern half of the country. Much of the country benefits from average wind speeds in excess of 5 m/s (often considered the minimum requirement for commercial viability), and significant areas are on a par with the best sites in Europe and the US with speeds over 7.5 m/s. The Figure 8.2 in the following shows the wind farms project in SA.

8.3 Hydrogen South Africa (HySA)—A National Hydrogen and Fuel Cell Technologies R&D and Innovation Strategy

SA is well-endowed with Platinum Group Metals (PGM, such as Platinum, Palladium, etc.) and other mineral resources such as Titanium, Vanadium, Chromium, and so on. Minerals and PGM beneficiation is currently a top priority for the SA government as it is hoped to unlock foreign investments and to attract downstream value-adding manufacturers, in turns creating jobs. The main focus is to add-value to titanium, iron, steel, and platinum. For the latter, the aim is to expand the Auto Catalyst sector and to create a Hydrogen and Fuel Cell industry in strategic geographical areas such as the Special Economic Zones (SEZs) [7, 8]. SA's mineral resources amounts to approximately USD$2.5 trillion and is the predominant supplier of PGM to the world, but not much beneficiation is currently undertaken in the country. However, the rise of Hydrogen Fuel Cell Technologies (HFCT) in various markets is about to change the global Platinum landscape with the anticipated increase in Platinum usage in this emerging market. Thus, it is safe to stipulate that if HFCT gains market share in coming years as is anticipated by the large vehicle manufacturers such as Toyota, Hyundai, Honda, and BMW, then the PGM market will see profound and sustained growth.

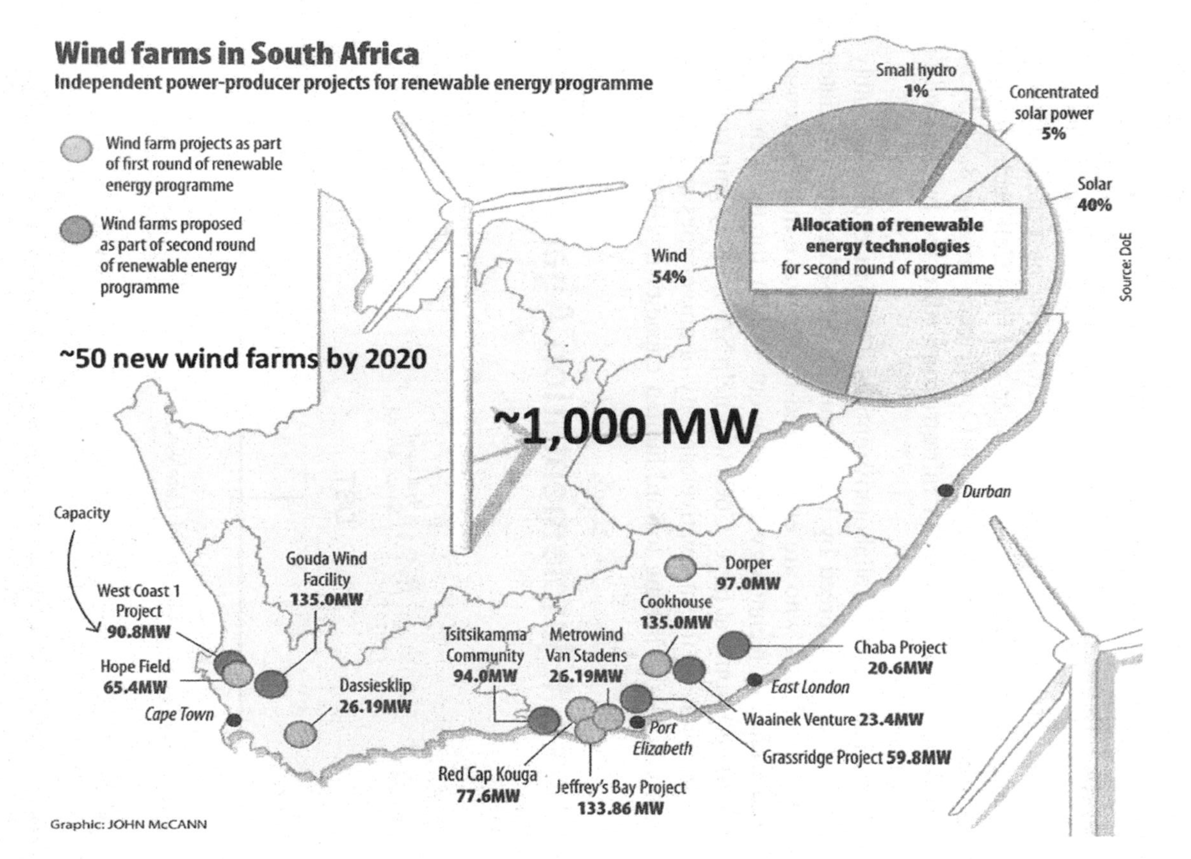

Figure 8.2 Map of current and future wind farms in SA—Graphic modified from [6].

In May 2007, Hydrogen South Africa or *HySA* was initiated by the SA Department of Science and Technology (DST) and approved by the Cabinet. *HySA* is a long-term (15-year) programme within their Research, Development, and Innovation (RDI) strategy, officially launched in September 2008. This National Flagship Programme is aimed at developing South African intellectual property, knowledge, human resources, products, components, and processes to support the South African participation in the nascent, but rapidly developing international platforms in Hydrogen and Fuel Cell Technologies. *HySA* comprises of three R&D Centers of Competence (Figure 8.3): *HySA* Catalysis, *HySA* Systems, and *HySA* Infrastructure (directed by Dr Dmitri Bessarabov at NWU) [7–14].

In SA, CoCs envisaged as collaborative entities established, and preferably led, by industry, and resourced by highly qualified researchers associated with research institutions who are empowered to undertake market-focused strategic research and technology development (RTD) for the benefit of industry and the economy at large. CoCs will help SA industry gain competitive advantage by using the innovative capacity of universities and research communities while contributing toward human capacity development. It is

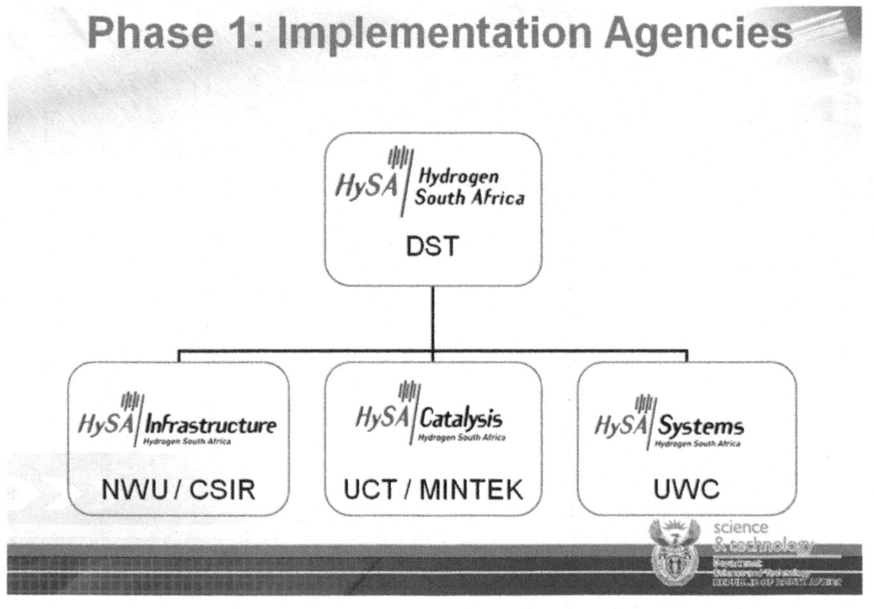

Figure 8.3 Phase 1—Establishment of a National R&D capability, comprising of three established centers of competence (CoC), based on a hub and spoke model.

envisaged that CoCs are formal, contractually secure, physical, or virtual platform upon which to establish collaborative technology development partnerships between government, industry, HEI, and SCs, with the explicit aim of technology commercialization.

The programme strives toward a knowledge-driven economy meaning that innovation will form the basis of SA's economy; this includes an aggressive capacity-development programme's approach. *HySA* also focusses on (i) the "Use and Displacement of Strategic Minerals", (ii) ways of harnessing South Africa's mineral endowments to promote both the Hydrogen Economy and renewable energy use, and (iii) seeking the most cost-effective and sustainable ways of incorporating PGM-based components in hydrogen fuel cell and other technologies, in turns resulting in commercialization ventures and a viable industry around mineral beneficiation.

HySA has been implemented in the context of the DST's various innovation strategies, the Department of Mineral Resources' minerals beneficiation strategy, the Department of Energy's Integrated Resource Plan (IRP, see earlier), and the Department of Trade and Industry's (DTI) industrial development strategies. The principal strategy of *HySA* is to execute research and development work, with the main aim of achieving an ambitious 25% share of the global Hydrogen and Fuel Cell market using novel Platinum Group Metal (PGM) catalysts, components, and systems since SA has more than 75% of the world's known PGM reserves. In order to achieve this, the structure is aimed at the parallel development of knowledge and technology across all areas of the Hydrogen and Fuel Cell value chain, allowing for the establishment of a strong R&D Hydrogen and Fuel Cell Technology exporting added value PGM materials, components, and complete products. Each Center has a unique responsibility, but all three are complementary within the common vision of fostering proactive innovation and developing the human resources required to undertake competitive R&D activities in the field of Hydrogen and Fuel Cell Technologies. The first five years of funding focused on developing infrastructures at each Center with a major emphasis upon Human Capacity Development (HCD). Relevant (inter)national expertise was recruited by each Center to access technical support and well-established implementation networks, and to ensure the programme and its deliverables remain market-related and world-class. Furthermore, to achieve the *HySA* strategy objectives, the three *HySA* Centers of Competence form a national network of research "Hubs" and "Spokes" through collaboration with institutions and partners from the R&D sector, higher education, as well as industry.

The DST published the 10-Year Innovation Plan for SA in 2007 [8]. According to this plan, one of the grand challenges' outcomes for SA to have achieved by 2018 includes:

- Energy security—the race is on for safe, clean, affordable, and reliable energy supply, and SA must meet its medium-term energy supply requirements while innovating for the long term in clean coal technologies, nuclear energy, renewable energy, and the promise of the "Hydrogen Economy".

SA's prospects for improved competitiveness and economic growth rely, to a great degree, on science and technology. The government's broad developmental mandate can ultimately be achieved only if SA takes further steps on the road to becoming a knowledge-based economy in which science and technology, information, and learning move to the center of economic activity. The knowledge-based economy rests on four interconnected, interdependent pillars:

- Innovation,
- Economic and institutional infrastructure,
- Information infrastructure,
- Education.

The DST of SA identified Hydrogen and Fuel Cell Technologies as one of its 'Frontier Science and Technology' initiatives. Hydrogen and Fuel Cell Technologies are widely seen as possible energy solution for the 21st century, yet it is far from clear how we will achieve what has been called the "Hydrogen Economy" in which the energy is stored and transported as hydrogen. At the core of fuel cell innovations is the platinum (Pt) catalyst. The emerging fuel cell market is expected to grow to a multi-billion dollar international industry. This places SA, which holds $\sim$87% of known Pt reserves, in a highly advantageous position. The DST is working to establish a specific policy framework to realize opportunities in these areas, along with a science and knowledge base that will ensure that SA benefits optimally from the nascent Hydrogen Economy. The primary focus for the establishment of the SA HFCT strategy is the beneficiation of natural resources, as articulated by the RSA Precious Metals Bill of 2005 [9].

One of the ambitious Grand challenge outcomes, according to the Ten-Year Innovation Plan for South Africa, includes:

- A 25% share of the global hydrogen infrastructure and fuel cell market with novel PGM catalysts;

- Have demonstrated, at pilot-scale, the production of hydrogen by water splitting, using either nuclear or solar power as the primary heat source.

It is clear that currently these goals serve as a high-level inspiration targets. The primary objectives of *HySA* include:

- Wealth creation through value-added manufacturing (this will be achieved by developing the platinum group metal (PGM) catalysis value chain in SA);
- Development of a hydrogen infrastructure (this will be achieved by developing local cost competitive hydrogen generation solutions based on renewable resources);
- Equity and inclusion in sharing the economic benefit derived from South Africa's mineral endowment (this will be achieved through creating a viable industry for the finished products that will create jobs and boost economic growth for the benefit of all South Africans); and
- Stimulation of PGM (in particular platinum) demand.

The key activities of *HySA* are covered in the following references [7–14].

8.4 Existing Industrial Hydrogen Infrastructure in South Africa

South African mining sector makes use of the large petrochemical operations producing hydrogen for refineries. Hydrogen is produced from natural gas and used by SASOL. However, SASOL made a provision to supply hydrogen for the base metal refinery at *Impala Platinum* mines. Installed hydrogen pipeline at *Implats*, longer than 100 km, can deliver up to 6 tonnes a day of excellent quality and purity hydrogen. Figure 8.4 shows hydrogen storage spheres at *Impala Platinum* [15].

8.5 Hydrogen and Fuel Cell Business Case in South Africa

8.5.1 Telecom Sector

Hydrogen systems have become widely known and are replacing batteries for back-up and Uninterrupted Power Systems (UPS) in applications where reliable back-up power is required. One such market is the telecommunications industry for emergency back-up power for mobile repeater stations in remote locations. These systems can be grid-connected and use fuel cell

Figure 8.4 Hydrogen storage spheres at Impala Platinum [15].

systems as emergency back-up or be autonomously powered by intermittent Renewable Energy (RE) with a hydrogen generator (PEM water electrolyser, ammonia reformer, etc) and fuel-cell system providing hydrogen storage for intermittent RE [16].

The high costs of battery technologies (especially lithium ion and redox flow batteries) and installation of electrical supply to remote locations in combination with the advances in renewable and hydrogen systems in the past few decades make renewable hydrogen systems a reliable and economical solution for remote telecommunication station electricity production.

In Africa, telecom towers are typically powered by diesel (generators), which often cause issues with diesel delivery, diesel cost and safety as well as thefts. An alternative to this is the use of fuel cell back up power systems. Sub-Saharan African telecom market is given in the following references:

- GSMA GPM market research and analysis: Powering Telecoms: East Africa Market Analysis, http://www.gsma.com/mobilefordevelopment/wp-content/uploads/2012/10/GPM-Market-Analysis-East-Africa-v3.pdf

- Powering Telecoms: West Africa Market Analysis, http://www.millennia2015.org/files/files/Zero_mothers_die/gpm_market_analysis_west_africa_.pdf

According to these reports, Sub-Saharan Africa mobile networks consist of more than 240,000 towers, providing only up to 70% coverage of population. It is expected that 325,160 towers will be operating by 2020. The majority of sites deployed so far are in off-grid or problematic areas (unreliable power supply). It is expected that by 2020, 189,000 towers will be operating at off-grid sites [17].

There have been a few FC telecom backup power and stationary baseload fuel cell installations in SA. Although it is well-accepted by the SA government that this technology has huge potential for the country – not only as a cleaner way to provide baseload electricity but also Pt is the main raw catalyst material in many fuel cell systems, which could increase demand for the country's enormous PGM reserves. It is also recognized that stationary baseload fuel cells have great potential for the country, both as a source of power and as an opportunity to establish a fuel cell manufacturing industry and value-added chain. In particular, they could play a key part in the growing market for distributed energy—or power that does not rely on large central generation plants and extensive grids. This is particularly attractive to parts of Africa where distributed energy could leapfrog power infrastructure bottlenecks.

Generally, stationary baseload fuel cells have a high capital cost, relatively low operating costs, a long life, high overall efficiency, and are readily available. When the additional heat produced by fuel cells is used for co-generation, their efficiency rates can be as high as 90% [18]. For example, the cost of running a mine in Africa on diesel using a diesel generator ranges between ZAR4–ZAR6 per kWh (at today's price). Depending upon the price of natural gas, the cost of running a fuel cell to power a mine could range between ZAR2–R3 per kWh. These figures indicate that on a total cost, or levelized cost, stationary baseload fuel cells compete well with the grid today for industrial and commercial customers in SA. According the IDC's estimates, the capital outlay for fuel cells could be about ZAR35 million to ZAR50 million per MW.

Some examples of demonstration projects are shown in the following:

- The Chamber of Mines in central Johannesburg is powered by a 100 kW platinum-based hydrogen fuel cell unit, fed by the *Egoli Gas* network. The unit powers a building that houses about 340 people, including businesses that rent spaces from the chamber. At peak power, the office block draws

120 kW of power although the whole building cannot be linked up to the fuel cell system because of its age and resultant complications with its wiring.

- Besides the Chamber's unit, which is based on Phosphoric Acid Fuel Cell (PAFC) technology, *Anglo Platinum* is using a fuel cell to power 34 homes, linked to a mini-grid, in the rural community of *Naledi Trust* near Kroonstad.
- *Impala Platinum* is aiming to use fuel cells to eventually take its PGM refinery in Springs off the national grid. The first phase of the project (phase I), which has just been completed (in 2016), will see 1.8 MW provided by fuel cells (*Fuji* PAFC systems running of waste hydrogen), with the excess heat that is produced being integrated into the refinery's operations. *Implats* uses the excess hydrogen it receives from industrial and speciality gas company *Air Products*. The second phase will see 22 MW of power supplied by fuel cells (running of natural gas, as well as direct hydrogen). Phase II of the project will comprise segments of 8 MW and 12 MW using 400 kW units with natural gas feedstock.
- A *Doosan Fuel Cell* PAFC system was installed in 2015 in collaboration with the South African DTI with a commitment to 1 GW of fuel cells to be installed by 2020.
- Cellphone giant *Vodacom* (a *Vodafone UK* subsidiary) uses fuel cells (an IdaTech ElectraGen^{TM} ME 5 kW system [*Ballard Fuel Cells*] including a fuel reformer that converts methanol/water liquid fuel into hydrogen gas to power the unit) running to power about 200 of its 10,000 baseload stations around the country. Here, the fuel cells used for backup power are at a capital cost of ZAR60 per kWh.
- In 2015, *PowerCell Sweden AB* and the power-as-a-service player *Mitochondria Energy Company (Pty) Ltd* signed a Letter of Intent to collaborate to develop diesel-fed fuel cell power solutions to the African market. PowerCell's power supply unit, the PowerPac, is currently under development. *Mitochnodria* will be involved in the pre-production development and testing processes to ensure the end product meets customer requirements.
- In 2012–2013, *HySA* Infrastructure Competence Center led by Dr Dmitri Bessarabov, director at the North-West University, launched for the first in Africa a Solar-to-Hydrogen Demonstration plant. The first generation commercial scale system consisted of PV modules (26 $\times$ 230 W) with Maximum Power Point Tracker (MPPT) (6 kW), tubular gel lead-acid

battery storage [640 Ah (C10) @ 48 V] with charge controller (2 × 5 W), PEM water electrolyzer (282 sL/h), high-pressure hydrogen storage cylinders (8 × 50 L), de-ionized water supply equipment, ventilation equipment, and a PLC and PC for control and monitoring purpose. A conceptual layout of the commercial scale system is given in Figure 8.5.

The commercial scale system capability has recently been upgraded from 6 kW PV to 15 kW PV, 30 kWh to 90 kWh battery storage, and 0.56 kg to 2.5 kg production of high-purity hydrogen per day. Additionally, an air-driven hydrogen booster provides filling pressures up to 200 barg with upgrade capability to 400 barg.

- In late 2014, *HySA* Systems Competence Center led by Professor Bruno G. Pollet (director, 2012–2015) at the University of the Western Cape launched a 2.5 kW Hydrogen Fuel Cell (HFC) Power Generator prototype unit to power the University's Nature Reserve. The fuel cell was sourced commercially and integrated by the researchers at the University. The system ran from a bank of hydrogen cylinders (*Air Products SA*).
- More recently, *HySA* Catalysis at the University of Cape Town reported the development of fuel cell catalysts (platinum supported on carbon) that could meet some of the industrial requirements for fuel cell stationary applications [19].

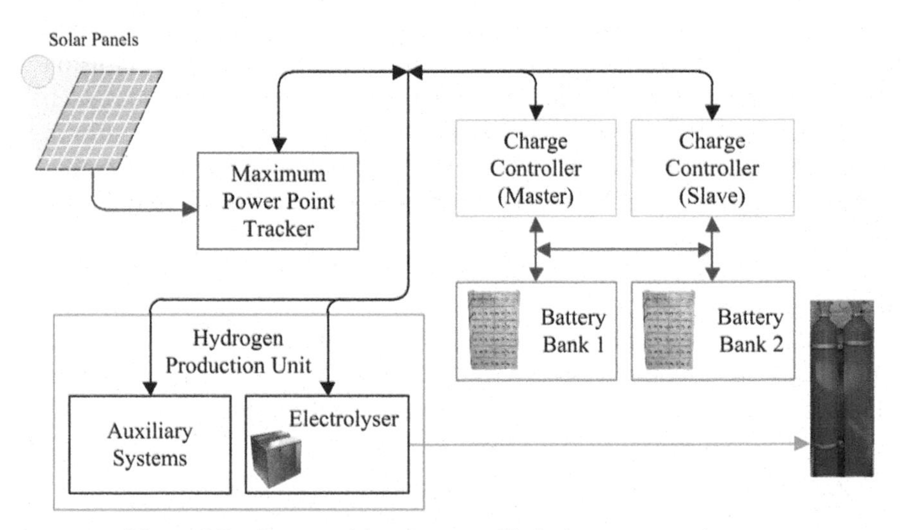

Figure 8.5 Commercial scale renewable hydrogen system layout.

However, hydrogen production and hydrogen distribution remain an issue in SA (see earlier). Depending upon the technology used, the feedstocks can vary from pure hydrogen (99.999%) to liquid petroleum gas, methanol, and biogas. In SA, telecommunication towers could particularly benefit from onsite hydrogen production, say for a 5 kW fuel cell backup power system connected to an electrolyzer and PV. For example, for a 5 kW fuel cell backup power system with an assumed 60% conversion efficiency, and assuming AC is required with DC to AC conversion efficiency of 92%, the power backup requirement is $[(5\ \text{kW} \times 8\ \text{hours})/(0.60 \times 0.92)] = 72.5\ \text{kWh}$. Using the Higher Heating Value (HHV) of hydrogen, $72.5\ \text{kWh}/39.4\ \text{kWh}\ \text{kg}^{-1} = 1.84$ kg of hydrogen would be required for 8-hour operation. For a telecommunication tower in the Johannesburg area (the mean sunlight hours in Johannesburg range between 7:25 per day in February and 9:42 to each day in August) with an average of 8 hours of sunlight per day, and assuming that the backup is designed for one 8 hour backup per month, 1.84 kg with 8 sunny hours, or 230 grams of hydrogen per sunny hour need to be produced.

8.5.2 Rural Electrification

In SA, around 1.3 million households (out of ~4.4 million) have no access to grid connection with ~600,000 rural homes where grid connection is uneconomical. Despite aggressive plans for expanding power capacity and transmission lines (Figure 8.6), it has been estimated that the grid extension would cost in the region of up to US$75,000 per km^2 with large areas such as the *Eastern Cape*, *Limpopo,* and *KwaZulu-Natal* set to still be off-grid in the future [5]. Currently, *Eskom* together the SA government are considering both grid extensions and off-grid technologies to electrify SA households by 2025 [5].

Table 8.1 shows the statistics for rural South African schools that have limited access to electricity [20].

Table 8.2 shows the electrification status of clinics in SA [21]. It is evident that there is a huge market opportunity for the deployment of fuel cell and local hydrogen production systems for these two applications.

Table 8.1 Number of rural schools without or limited access to electricity [20]

	Total	No Electricity	Unreliable Electricity
# Schools in RSA	24,793	3,544	804

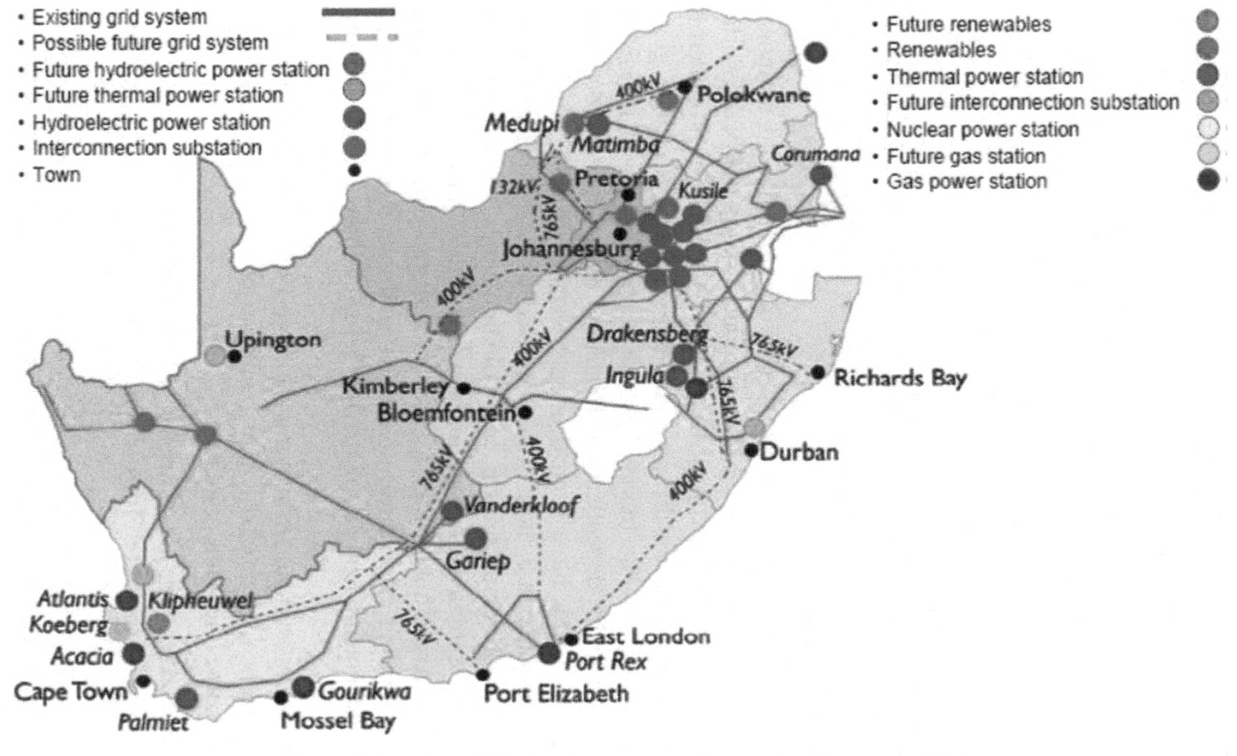

Figure 8.6 Map of SA showing power stations and grid systems [5].

Table 8.2 Electrification status of clinics [21]

				No Electricity	
	Total	Urban	Rural	Major Clinics	Day Clinics
Eastern Cape	450	249	201	317	40
Limpopo	400	48	352	282	10
KwaZulu-Natal	348	271	77	77	31
Gauteng	430	428	2	0	0
Free State	240	177	63	5	7
Mpumalanga	220	95	125	100	7
North West	290	127	163	125	8
Northern Cape	120	94	26	5	20
Western Cape	450	428	22	0	25
Total	2,948	1,917	1,031	911	148

8.5.3 Mining Sector

Africa hosts the world's largest mineral reserves of platinum, gold, diamonds, chromite, manganese, and vanadium. It also produces about 17% of the world's uranium. The continent's growth in the past has mostly been propelled by exaggeratedly high prices for these commodities. Mining is a huge global industry, but it faces major stresses because it provides raw materials for economic growth, highly dependent on global outlook for the world economy. Revenues in this sector have been dropping (peaked in 2008) and therefore mining companies are increasingly focused on cost containment and managed production. One area for cost containment is in the energy use as mining is an extremely energy-intensive industry. Australia, Canada, Latin America (Argentina, Chile, Mexico, and Peru), and SA are major targets of new mining exploration investment and thus represent appropriate countries of potential intake of the HFC Technologies, *know-how*, and so on.

Mining in SA is estimated to account for about 18% of the country's GDP, 20% of investment, and approximately 1 million jobs (500,000 direct). The following are some important points:

- Total mineral reserves are estimated at US$2.5 trillion.
- The industry has suffered from rising energy prices in the past several years.
- Although government officials have regularly said nationalization is not on the table, the issues continues to be debated, especially with continuing layoffs and closures.

Currently, the mining sector faces a number of challenges such as high operating costs associated with ventilation costs, maintenance of diesel equipment, and lack of automation and tele-remote operations.

WHO classifies diesel engine exhaust as "carcinogenic to humans," based on sufficient evidence that it is linked to an increased risk of lung cancer and lung-related diseases, thus underground mine workers are at high risk because the machines they use are diesel-powered. The cost of the flame-proofing of diesel equipment is also high. The safety aspects and codes for hydrogen application in the mining environment need further development and validations. The requirements for hydrogen infrastructure need to be better understood.

Potential Drivers and Challenges for Fuel Cell Use in Underground Mining Vehicles—Mining revenues have dropped over the past several years, and mines are facing economic pressure. Mining is an energy-intensive industry, and energy reduction is a target for mining companies looking to control costs. The amount of diesel fuel required for moving a fleet of excavation and hauling vehicles 24 hours per day makes mines vulnerable to fossil fuel price fluctuations. Underground mining requires managing heat from diesel engines and ventilation, both of which add substantial cost. Diesel Particulate Matter (DPM) is a health hazard. While diesel engines are becoming cleaner, this will increase costs and can reduce efficiency, which adds to fuel costs; these factors can reduce the cost barrier to adopting alternative technologies such as fuel cells.

Fuel cells are becoming an economically viable option as a power source for back-up power, transportation, and so on. This technology shows considerable benefits and potentially to be used in underground mines, for which the onsite production and risks around the safety of supply of hydrogen to the fuel cells and on-board storage need to be assessed. The underground mining environment is perfect for hydrogen fuel-cell implementation because of the very strict safety regulations that is applied in underground mining operation. In terms of storage, various storage technologies exist. These include compressed storage (up to 350 barg), liquid storage, metal hydride storage, and liquid organic hydrogen storage. Compressed hydrogen is the most widely implemented and by far the most mature technology with the interest from motor vehicle manufacturers causing a large capital input into research and development of safety technologies for on-board operation that will be of benefit for underground implementation. Liquid hydrogen storage is also a mature technology and used widely for bulk storage but is only viable for above ground storage applications due to the nature of liquid storage. Metal hydrides are in pre-commercial stage with very low energy densities by weight being a drawback and development phase to improve on the energy densities. Liquid organic hydrogen storage is still in the beginning of the research phase, but shows good promise with relative high-energy densities.

Reliable infrastructure that produces hydrogen onsite and deliver it to mining operations is required in order to have the ability to test hydrogen-based applications for the mining sector. It is an example of the "chicken-and-egg" situation where safety concerns have to be addressed and cost-effective infrastructure for the production of hydrogen is required in order to deploy fuel cell applications. Using hydrogen commercially in underground mining applications is so novel at this stage that the Canadian and US governments and global players have teamed up to develop safety standards for it.

The use of hydrogen as a fuel inevitably results in risks to the public as is the case for use of any combustible materials. Currently, hydrogen utilization is predominantly limited to highly-trained individuals. If it is to be used in public, untrained people must be able to handle hydrogen with the same degree of confidence and with no more risk than conventional liquid and gaseous fuels. Risk should be regarded as the product of the probability of an incident or accident occurring and the magnitude of its hazardous consequences. It is therefore evident that the reliable, cost-effective method of producing hydrogen on-site (without long-distance delivery) is required for demonstration and training before it can be used underground.

Fire and explosion hazards must be carefully assessed to determine the relative safety of a fuel for each potential application. Hydrogen can be safer than conventional fuels in some situations and more hazardous in others. The relative safety of hydrogen compared to other fuels must be taken into consideration in the particular circumstances of its accidental release. Several reviews [22–25] have been published that consider the safety of hydrogen as a vehicle fuel. These have concentrated primarily on hydrogen safety related to the vehicle itself rather than the wider context of a fuelling infrastructure and especially, for use of hydrogen in mining environment underground.

Little information on the behavior of hydrogen (release, turbulence) in enclosed spaces (none for mining) with/without ventilation and behavior versus Lower Flammable Limit (LFL) is available, such as design, performance, and reliability of on-board hydrogen storage: safety assessment and hazard analysis. Currently, mine regulations for hydrogen codes and standards are not available for using hydrogen PEM fuel cell power and storage. It is of paramount importance that site hydrogen storage and distribution infrastructure needs to be established to initiate the deployment of mining fuel cell vehicles.

Overall, it will establish a platform for a large project that potentially can grow into the mega project category, (i) the level of localization of technology solutions to address local and international problems, (ii) the potential market capitalization and its economic impact to SA GDP, job creations, IP development, and (iii) addressing health issues. Mining companies face strong pressures to adopt more sustainable practices. Mining companies are also focusing on cost cutting, and energy use is a key cost area for this effort. Some alternative fuel technologies are being adopted in vehicles, especially natural gas and EVs. Underground mining currently offers the best opportunity for fuel cell integration into vehicles due to the safety, health, and cost concerns surrounding these vehicles' emissions and the need for ventilation.

To date, there has been little activity to develop fuel cell for mining vehicles; these applications require cost reduction and improvements in hydrogen supply and storage. The best opportunities for fuel cells in mining vehicles would be countries like Canada and SA, which have a focus both on promoting fuel cell technology and also on sustainability or good corporate stewardship, as well as Australia, which has made sustainability a key goal for its mining sector [26].

8.5.4 Power-to-Gas in South Africa

Around the world, the development of clean sources of power generation has increased rapidly over the past thirty years. In many jurisdictions today, such as Germany and California (USA), over 20% of the electricity consumed comes from renewable energy sources, predominately wind and solar. The challenge with these renewable sources is that they are intermittent. Unlike a thermal power plant such as a gas-turbine-driven power plant, it is not possible to dispatch the wind or the sun and turn it on to match the demand profile of when electricity is needed. When renewable generation makes up a small portion of the generation grid mix, the task of integrating their output is relatively simple.

However, when the proportion of renewable generation reaches a critical mass, there may be frequent periods during the day when demand is lower than production, and even consecutive days during the late spring and early fall where there is surplus power generation. Today, the default solution is simply to curtail or waste the wind or solar generation. This is truly a waste of a resource. Although the electrical power grid allows for easy distribution of energy, the grid itself has no storage capacity. To store electricity, it usually has to be converted into other forms of energy. A variety of storage technologies currently exist such as electrical energy storage (super capacitors), potential energy storage (traditional pumped storage), mechanical energy storage (compressed air reservoirs, flywheels), electrochemical energy storage (batteries), and finally chemical energy storage (hydrogen, synthetic natural gas, methanol). The rated power, energy storage capacity, charge and discharge time and frequency, capital and operation costs characteristics of these storage systems vary greatly.

A successful transition toward a cleaner and more sustainable energy system in 2050 requires a large-scale implementation of sustainable and renewable energy sources. The European CO_2 emission reduction target of 80% in 2050, relative to 1990 emission level, implies that the power production sector should be fully sustainable by then and that other sectors, like the industry and mobility sector, should rely largely on the sustainable use of energy sources. The implementation of energy saving measures as well as an adequate selection of power resources (e.g., renewable as well as low-carbon) is necessary.

Power to Gas (P2G) is a technology, which converts electrical power to a gas fuel. There are two methods, the first one is to use the electricity for water splitting by means of electrolysis and inject the resulting hydrogen into the natural gas grid. The second method is to combine the hydrogen with carbon dioxide (CO_2) and convert the two gases to methane (CH_4), using electrolysis and a methanization reaction. The excess power or off peak power generated by wind generators or solar plants is then used for load balancing in the energy grid. P2G is a technological concept that enables controllable power demand load and offers the opportunity for (i) electricity storage if hydrogen is converted back to electricity, (ii) accommodation in the gas infrastructure, either by direct injection of hydrogen or by the conversion of hydrogen and carbon dioxide into methane by the *Sabatier* process, (iii) application of hydrogen as feedstock in the industry, and (iv) application of hydrogen as fuel in the mobility sector.

Hydrogen production from renewables in SA is shown to be successful with technologies and experts capable of meeting any power-to-gas requirements. From the P2G technology options, chemical methanization, biological methanization, industrial hydrogen feed, and hydrogen fuelling are viable technologies that can be considered for SA.

Storage in the natural gas network might be an option in future with the South African government gas utilization master plan investigating the

Commercial Platform for Power-to-Gas in Africa

Products	Clients: CNG consumers	Clients: SASOL	Clients: ESKOM	Clients: Egoli gas	Clients: CNG group of companies	Clients: Waste water treatment works	Clients: Waste disposal facilities	Benefits: Waste mamangement	Benefits: CO2 Reduction	Benefits: Reduce cost of fuel	Benefits: Reduce cost of energy	Benefits: Reduce load on ESKOM
Methane production	X	X	X	X	X				X	X	X	X
Biogas enhancement	X	X	X	X	X	X	X	X	X	X	X	X
Hydrogen production		X	X	X					X			X
Hydrogen/methane storage				X	X	X	X		X			X

Figure 8.7 Platform for power-to-gas in Africa [28].

infrastructure necessary to open up gas prospect. The master plan investigates the integration with the gas finds in *Mozambique* and the *Karoo* (RSA). Further consideration is the conversion of the diesel-fuelled open-cycle gas turbines that are either operational or under development in *Saldanha Bay*, *Mossel Bay*, *Coega* in Port Elizabeth, Durban, and *Richards Bay* to closed-cycle gas turbines fuelled using gas [27].

Figure 8.7 provides technologies, potential clients, and benefits of a power-to-gas platform in SA.

8.6 Prospects and Conclusions

Fuel Cells and associated hydrogen Infrastructure represent an exciting new market, which could drive growth for platinum as well as spark significant new opportunities internationally and locally in SA. Hydrogen and Fuel Cell Technologies for the mining and telecom sectors as well as for rural electrification could potentially create new large market opportunity for fuel cell technology deployment in SA and internationally. The following are the benefits of developing hydrogen infrastructure and fuel cell market in SA:

- Means of meeting increasing demand for energy,
- Reduction of carbon footprint,
- Platform for mineral beneficiation,
- Opportunity for job creation,
- Wealth creation,
- Export opportunities,
- Increase demand for Platinum Group Metals.

SA needs to adopt a different strategy with regard to power generation, that is, to produce energy locally in order to sustain its long-term growth, ambitions, and success. This is due to the fact that investing in power generation infrastructure in the short term will add significantly to manufacturing capital costs and thus will impact on competitiveness and economic growth.

Finally, the authors are inviting the reader to consult the latest report by the *Mapungubwe Institute for Strategic Reflection* highlighting the strategic role of Platinum Group Metals on SA and the global Hydrogen Economy [29].

References

[1] IRENA, 2015, "Prospects for the African Power Sector". https://www.irena.org/DocumentDownloads/Publications/Prospects_for_the_African_PowerSector.pdf

[2] Kabiru, I. M. Does clean energy contribute to economic growth? Evidence from Nigeria. Energy Reports. 2015; 1: 145–150.

[3] Pollet, B. G., Staffell, L., Adamson, K. A. The current energy landscape in the republic of South Africa. International Journal of Hydrogen Energy. 2015; 40: 16685–16701.

[4] Presentation by Pollet, B. G. Energy Landscape in Emerging Economies: A South African Perspective, HFC2013, June 16th–19th 2013, Vancouver.

[5] Reference: Roland Berger Feasibility Report, June 2012 in Presentation by Anglo American entitled: Rural Electrification: Fuel Cell Power Systems, November 2014.

[6] Map created by McCann, J. Cape Angus Newspaper, 2013.

[7] Pollet, B. G., Pasupathi, S., Swart, G., Mouton, K., Lototskyy, M., Williams, M., Bujlo, P., Ji, S., Bladergroen, B. J., Linkov, V. Hydrogen

South Africa (HySA) systems competence center: mission, objectives, technological achievements and breakthroughs, International. Journal of Hydrogen Energy. 2014; 39(8): 3577–3596.
[8] Innovation towards a Knowledge – Based Economy, Ten-Year Plan for South Africa, 2008–2018, www.esastap.org.za/download/sa_ten_year_innovation_plan.pdf
[9] Precious Metals Bill, RSA Government Gazette No. 27929, 19 August 2005, http://www.gov.za/sites/www.gov.za/files/b30-05_0.pdf
[10] Bessarabov, D., van Niekerk, Frik., van der Merwe, F., Vosloo, M., North, B., Mathe, M. Hydrogen infrastructure within HySA national program in South Africa: road map and specific needs. Energy Procedia. 2012; 29: 42–52.
[11] Bessarabov, D., Human, G., Chiuta, S., van Niekerk, F., de Beer, D., Malan, H., Grobler, L. J., Langmi, H., North, B., Mudlay, D., Mathe, M. HySA Infrastructure Center of Competence: A Strategic Collaboration Platform for Renewable Hydrogen Production and Storage for Fuel Cell Telecom Applications, 2014 IEEE International Telecommunications Energy Conference, (INTELEC 2014), Vancouver, Canada, 28 September–2 October 2014.
[12] Barrett, S., Bessarabov, D. HySA infrastructure: producing and using hydrogen for energy in South Africa. Fuel Cells Bulletin. 2013; 12–17.
[13] Barrett, S., Pollet, B. G. Mission and objectives of the Hydrogen South Africa (HySA) systems competence center. Fuel Cells Bulletin. 2013; 10–17.
[14] Barrett, S., Blair, S. HySA/catalysis: creating opportunities from South Africa's mineral wealth. Fuel Cells Bulletin. 2013; 12–15.
[15] Presented by *Impala Platinum* at CARISMA2014, December 1st–3rd 2014, South Africa.
[16] Varkaraki, E., Lymberopoulos, N., Zachariou, A. Hydrogen-based emergency back-up system for telecommunication applications. Journal of Power Sources. 2003; 118: 14–22.
[17] Presentation by Bessarabov, D. Development and Deployment of Hydrogen and Fuel Cell Technologies in South Africa: Prospects and Challenges, International Renewable Energy Storage Conference (IRES), 09–11 March, Dusseldorf, Germany, 2015.
[18] Ellamla, H. R., Staffell, L., Bujlo, P., Pollet, B. G., Pasupathi, S. Current status of fuel cell-based combined heat and power systems for residential sector., Journal of Power Sources. 2015; 293: 312–328.

[19] van Schalkwyk, F., Pattrick, G., Olivier, J., Conrad, O., Blair, S. Fuel Cells, 2016, DOI: 10.1002/fuce.201500161
[20] NEIMS (National Education Infrastructure Management System) Reports, 2011 (https://edulibpretoria.files.wordpress.com/2008/01/school-infrastructure-report-2011)
[21] Renewable energy world, March–April 2002 (http://energy4africa.net/klunne/publications/pv_schools_clinics_rsa_rew.pdf)
[22] Directed Technologies Inc. Direct-hydrogen-fuelled proton-exchange-membrane fuel cell system for transportation applications, Hydrogen safety report, DOE/CE/50389-502, May 1997.
[23] Barbir, F. Safety issues of hydrogen in vehicles, International Association for Hydrogen Energy, Technical Papers, http://www.iahe.org/hydrogen%20safety%20issues.htm
[24] Cadwallader, L. C., Herring, J. S. Safety issues with hydrogen as a vehicle fuel, INEEL/EXT-99-00522, 1999.
[25] Ringland, J. Safety issues for hydrogen-powered vehicles, Sandia National Laboratories, SAND94-8226, March 1994.
[26] Presentation by Bessarabov, D. Hydrogen and Fuel Cells for Mining Equipment, The World Hydrogen Technologies Convention (WHTC), Sydney, Australia, October 11–14, 2015.
[27] http://africaoilgasreport.com/2014/07/gas-monetization/eskom-seeks-gas-assets-to-rein-in-excessive-diesel-costs/
[28] Bessarabov, D., Human, G. Solar Energy for Hydrogen Production: Experience and Application in South Africa, SASEC 2015, Third Southern African Solar Energy Conference, 11–13 May, 2015, Kruger National Park, South Africa.
[29] South Africa and the Global Hydrogen Economy, the Strategic Role of Platinum Group Metals, Mapungubwe Institute for Strategic Reflection, MISTRA publication, 2013, ISBN 978-1-920655-68-6, 400 pages.

PART III

European Context

9

Hydrogen Energy Transition in European Union

Ioan Iordache[1,2] and Ioan Ştefănescu[1,2]

[1]National Research and Development Institute for Cryogenics and Isotopic Technologies ICSI, Rm. Valcea, Romania
[2]Romanian Association for Hydrogen Energy, Rm. Valcea, Romania

9.1 Introduction

In order to understanding the progress made by hydrogen in Europe, everyone must understand the context of the world progresses and the environment and climate circumstances of this planet. In the next words, three recent benchmarks are disclosed. COP 21 in Paris, in the previous year, was an event which became one of the major turning points in our relationship with the environment and climate. The Climate Conference's chief aim was to achieve a legally binding and universal agreement on climate, with the aim of keeping global warming below 2°C, for the first time in over 20 years of United Nations negotiations. In the 2030 Agenda for Sustainable Development, after negotiations and debate, 193 countries had agreed to set 17 sustainable development goals more bold and ambitious than anything that has come before them, United Nations Headquarters, New York, September 2015. The 17 Sustainable Development Goals and 169 targets will stimulate action over the next fifteen years in areas of critical importance for humanity and the planet and will demonstrate the scale and ambition of this new universal Agenda. Following the adoption of Directive 2014/94/EU on the deployment of alternative fuels infrastructure, the European Commission (EC) created the Sustainable Transport Forum (STF), April 2015. The forum envisages, in particular, to: (1) provide advice and technical expertise on the development and implementation of the alternative transport fuels and contribute toward an energy-efficient, decarbonized transport sector; (2) facilitate the exchange of information on initiatives, projects, and partnerships dealing with alternative transport fuels; and (3) deliver opinions, submit reports, or develop and propose innovative solutions on any matter of relevance to the promotion of alternative transport fuels in the EU.

In their recent pages, Pierre-Etienne Franc and Pascal Mateo consider hydrogen a decisive issue for Europe, a vital matter, with economic, geopolitical, and societal ramifications, all concatenated as a question of civilization. These remarkable remarks may be shaped like a hydrogen pyramid, which summarize the four issues. From the economic perspective, the debottlenecking access to clean and renewable energy would create a wealth of jobs and technologies required for the continuous development. On the geopolitical level, the Europe will be able to develop various forms of local autonomy in terms of both access to and use of renewable energy sources. The Europeans would escape from political constraints caused by the power relationships between countries that produce and those that consume oil and gas resources. From a societal perspective, the restructuration of the energy industry, energy for mobility also is included, would be a massive source of the reshoring and secure of jobs and expertise. Finally, a question of civilization that our civilization was built on a base of individual liberty, the abundance of resources, and technical progress, but was captive in a war of dwindling resources between societies and classes. There is need of a future that is polar opposite and the first step toward it is to launch the energy transition [1].

9.2 SET-Plan

The European Union (EU) has globally positioned itself with strong targets on CO_2 emissions reductions, energy efficiency, and the use of renewable energy resources. The EU drivers for energy research, development, and innovation are mitigations of climate exchange, warranting security of supply and economic-industrial development. To support this goal the European stakeholders have established a series of concrete actions. In its SET-Plan (Strategic Energy Technology Plan), EU has established seven technology focus areas: wind energy, solar energy, carbon capture and storage, nuclear energy, bio-energy, electricity grids, finally but not the last, fuel cell and hydrogen. At the EU level is observed a huge increase in public budgets for the activities referring to renewable energy and energy efficiency research, development and demonstration while the budgets for traditional energy sources are not changing significantly [2].

The objectives of the European energy research programs, Member States and EU, are to aid the creation and establishment of the technology necessary to move the current energy system toward one that is more sustainable, competitive, and secure. The next energy system must use a more diverse energy mix of energy sources in special renewables and non-pollutes and new energy carriers. The EU supports energy research over a broad portfolio of technologies and its pillar is SET-Plan. That was adopted by the EC in 2008 with the aim of putting in the place an energy technology policy for Europe. The principal objective of this initiative is to enhance the development of low-carbon energy technologies leading to their market uptake while boosting Europe's competitiveness in the field. The SET-Plan is not only a vehicle to promote renewable energy resources but it also contributes to two of the seven Flagships Initiatives of the EU, "Innovative Union" and "Resource Efficient Europe".

To take account of the large variety of energy research topics, the Energy Research Knowledge Center (ERKC) has structured the various themes from the desire of help achieve the SET-Plan objectives. The thematic structure has been categorized into 45 themes covering 9 priority areas. Hydrogen and fuel cell was included into priority area number two, alternative fuels and energy resources for transport.

Under the umbrella of SET-Plan are found also the SET-Plan Information System (SETIS), the European Energy Research Alliance (EERA), and the SET-Plan Steering Group. SETIS is the EC's Informative System for SET-Plan, managed by the Joint Research Center (JRC). EERA was founded by European research institutes as a non-profit organization established in Bruxelles. The Steering Group consists of high-level representatives from the Member States and the EC.

There are the following three key components of SET-Plan implementation: (1) The Smart Cities and Communities European Innovative Partnership that will sustain the demonstration of energy, transport, and information and communication technologies in urban areas for cities' needs. (2) European Energy Research Alliance Joint Programmes (EERAJPs), the aim of it is to gradually evolve into fully operational, virtual research institutes. Sixteen Joint Programmes have been established in a wide range of energy research fields. One of them is Joint Programme on Fuel Cells and Hydrogen, it aims to accelerate and harmonize the long-term research on fuel cells and electrolyzers in Europe, and summarizes six structures:

Electrolytes, Catalysts and Electrodes, Stack Materials and Design, Systems, Modeling-Validation-Diagnosis, and Hydrogen Production. The European Energy Research Alliance (EERA) has been working since 2008 to align the research and development activities of individual research organizations to the needs of the SET-Plan priorities, and to establish a joint programming framework at the EU level. (3) Joint large-scale technology development projects where included are the European Industrial Initiatives (EIIs): Wind, Solar, Carbon capture and storage, Electricity grids, Bio-energy and Nuclear fission, and one Fuel Cells and Hydrogen Joint Undertaking (FCH JU), Figure 9.1.

Figure 9.1 SET-Plan and Fuel Cells and Hydrogen Joint Technology Initiative.
Source: FCH JU.

9.3 Fuel Cells and Hydrogen Joint Undertaking

Under the Horizon 2020 the bulk of funds for fuel cells and hydrogen research will be entrusted to the Joint Technology Initiative (JTI). FCH JU is the legal entity entrusted with the coordinated use and efficient management of the funds committed to the JTI. It is also a public-private partnership with the following three members: the EU, represented by the EC; the industry grouping named Hydrogen Europe; and the N.ERGHY Research Grouping.

There is a strong increasing trend in EU funding to fuel cells and hydrogen research over successive FPs. Under FP2 (1986–1990) it was 8 M Euro; by FP3 (1990–1994) it was 32 M Euro; by FP4 (1999–1998) it was 58 M Euro; by FP5 (1999–2002) it was 145 M Euro and in FP6 (2002–2006) it was around 320 M Euro, matched by an equivalent amount of participating stakeholder investment. In FP7, when FCH JU effective began to run, was provided a budget of 470 M Euro and the private sector will contribute matching funds of at least 470 M Euro. Under the new arrangements, Horizon 2020 will provide a budget of 665 M Euro and the private sector will contribute matching funds of at least 665 M Euro.

The next paragraphs will disclose the "history" of FCH JU. The EC's High Level Group presented a report in 2003 entitled "Hydrogen Energy and Fuel Cells—a vision of our future" that recommended the formation of a partnership between the EC and industry for the development of fuel cell and

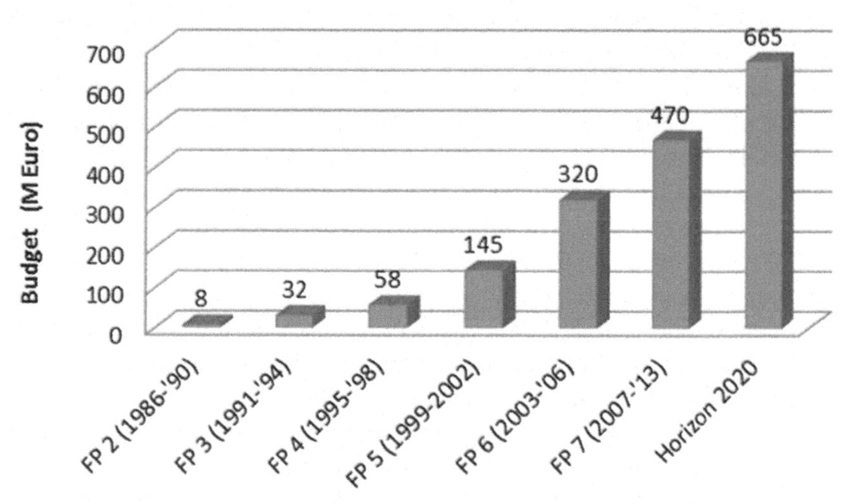

Figure 9.2 EC funds to Hydrogen and Fuel Cell research in the various FPs.

hydrogen technologies. The report also recommended the establishment of a pilot programme, with EC's funding, to make those technologies commercially viable. In the same year, the EC adopted its "European Initiative for Growth" programme that established a hydrogen economy project with a budget of 2.8 bn. euros for the decade 2004 through 2015. The initiative allowed for possible funding from the structural funds and the EC's "Research, Technological development and Demonstration Framework Programmes." The initiative placed an emphasis on the long-term research and cooperation with advisory boards. After that the EC sustained the establishment of a platform named "European Hydrogen and Fuel Cell Technology Platform" (HFP) that put together interested partners in a joint venture. The platform adopted a research agenda for accelerating the development and market introduction of fuel cell and hydrogen technologies within the Europe in 2005. This agenda called for funding by the EC and organizations from the public and private sector.

In 2006, the agenda of the technology platform was adopted by the Council Decision 2006/975/EC within the EC's Seventh Framework Programme. The prospect of further financing from the European Investment Bank (in particular through its Risk-Sharing Finance Facility) had been established in an earlier decision (2006/971/EC). In 2007, the European Council concluded that the Member States had an interest in renewable energy programs. On 10 October 2007, the European Commission establishment up of the "Fuel Cells and Hydrogen Joint Technology Initiative" (JTI). This proposal was accepted by the European Parliament and the Council of Ministers, and on 30 May 2008 the Council passed regulation number 521/2008 setting up the "FCH JU" that will run until 31 December 2017. Article 2 of the regulation stipulated a contribution to the implementation of the Seventh Framework Programme, in particular to its energy-, nanotechnologies-, environment-, and transport-specific programmes. The "Fuel Cells and Hydrogen Joint Technology Initiative" was officially launched on 14 October 2008 during the General Assembly of Fuel Cells and Hydrogen Stakeholders.

The strategic approach in FP6 was to support the selected fuel cells and hydrogen technologies across the spectrum of research and development, from basic research to demonstration projects, complemented by projects on cross-cutting issues. Around 75% of the hydrogen and fuel cell research projects was funded under the thematic priority on "Sustainable Energy Systems" of FP6, but other thematic priorities and FP6 programmes have also contributed [3]. Starting with FP7 and now with Horizon 2020, the financed projects are sensitive and focused on demonstration and commercialization of hydrogen and fuel cell technologies. The FCH JU provide financial support mainly in a form of grants to participants following open and competitive calls for proposals. Other instruments foreseen in Horizon 2020 also is used for studies through a public procurement procedure.

The aim of the FCH JU under FP7 was to accelerate the development and deployment of fuel cells and hydrogen technologies by executing an integrated European programme of RTD activities for the period 2007–2013. The new phase of the FCH JU under Horizon 2020 will build on the experience gained in the current period 2008–2013 in order to allow for a leaner governance structure, involvement of a broader range of stakeholders, more efficient operations, and an optimal management of the human and financial resources. In particular, the FCH JU will aim not only at a better alignment and coherence of the national, regional, and JU programmes but also at fostering jointly funded actions, smart specialization in regions, and the complementary use of Structural Funds [4].

The FCH JU programme, with an operational budget nearly 1 bn. euro, jointly contributed 50/50 by public and private partners, has served as a key growth catalyst for hydrogen and fuel cell in Europe. The programme united the various players in the sector and provided predictability. This long-term commitment offers a stable framework for the research, development, and demonstration activities, which otherwise would have been impossible in difficult economic times. The programme helps stakeholders build coalitions that act as a central focal point. The FCH JU was in the situation to put the individual players together into a successful manner and links between national initiatives. Through its Annual and Multiannual Implementation Plans, FCH JU developed a joint coordinated strategy to increase effectiveness. The FCH JU funding mechanism pooling resources to support nascent technologies beyond local or private possibilities.

As a result of FCH JU support, the hydrogen and fuel cell sector has grown substantially: 10% average increase in annual turnover, 500 M Euro in 2012, 8% average increase of research and development expenditures, 1,800 M Euro in 2012, 6% average increase of market deployment expenditures, 600 M Euro in 2012, 6% growth in jobs per year while the EU job market has contracted, and 16% annual increase in patents [5]. The most recent programme, 2008–2014, has provided a strong and stable growth platform for Small and Medium Enterprises (SMEs), who has considered valued partners of hydrogen and fuel cell community. The SME participation rate in hydrogen and fuel cell-related projects was 25.6%, higher than that FP7 average rate of 18%. The focus of FCH JU projects, achieving substantial progress in both energy and transport, ranges from basic research to large-scale demonstration and premarket studies [6]. The next paragraphs contain general information regards references on thematic of projects according with the same bibliographic source.

The main objective of the transport and refuelling infrastructure pillar was the development and testing of competitive hydrogen fuelled fuel cell electrical vehicles (FCEVs), the corresponding refuelling infrastructure, and the full range of supporting elements for market development, details are available in the Figure 9.3 and as well as on FCH JU website. The key achievements of the projects are listed: 49 fuel cell and hydrogen buses, 37 passenger fuel cell electrical vehicles, and 95 mini cars with range extenders on the European roads; 13 new hydrogen refuelling station with an availability of 98%; the cost of hydrogen "at the pump" under 10 Euro per kg; and reduction of hydrogen consumption in buses from 22 to less than 11 kg/100 km. The longest and biggest project is *CHIC* and brings together 25 partners, will operate 26 fuel cell buses in 5 cities with a total of 10 fuelling stations across Europe. The cost of the projects was 82 M Euro, only 26 M Euro funding from FCH JU, the period of the projects is between 2010 and 2016.

The second pillar, regarding hydrogen production and distribution, aimed to develop a portfolio of cost-competitive, energy efficient, and sustainable hydrogen production, storage and distribution processes. These efforts were necessary to demonstrate the role of hydrogen as an energy carrier in reaching Europe's key energy objectives: renewable production of hydrogen and decarbonization, details are available in the Figure 9.4. The key targets of the projects are: new materials for the removal of asbestos in alkaline electrolyzers, large (MW) scale PEM electrolyzers, power electronics for connecting intermittent renewables, electrochemical compression of hydrogen, energy consumption optimization and safety.

The overall objective of the stationary power generation and combined heat and power (CHP) application pillar was to enhance the quality of current technology by pass over the gap between laboratory and pilot prototypes and pre-commercial systems. That means product validation under real market conditions, scaling up the manufacture capacities for industrial production, details are available in the Figure 9.5. The support provided by FCH JU was highly application orientated and technology neutral. The research aspects were focused on cost reduction through new materials and technologies, longer lifetime through understanding of degradation issues, diagnostic and control tools. The concrete achievements are more than 1,000 micro CHP addressing to market for domestic applications, electric efficiency up to 60% for SOFC and below 20,000 Euro/kW.

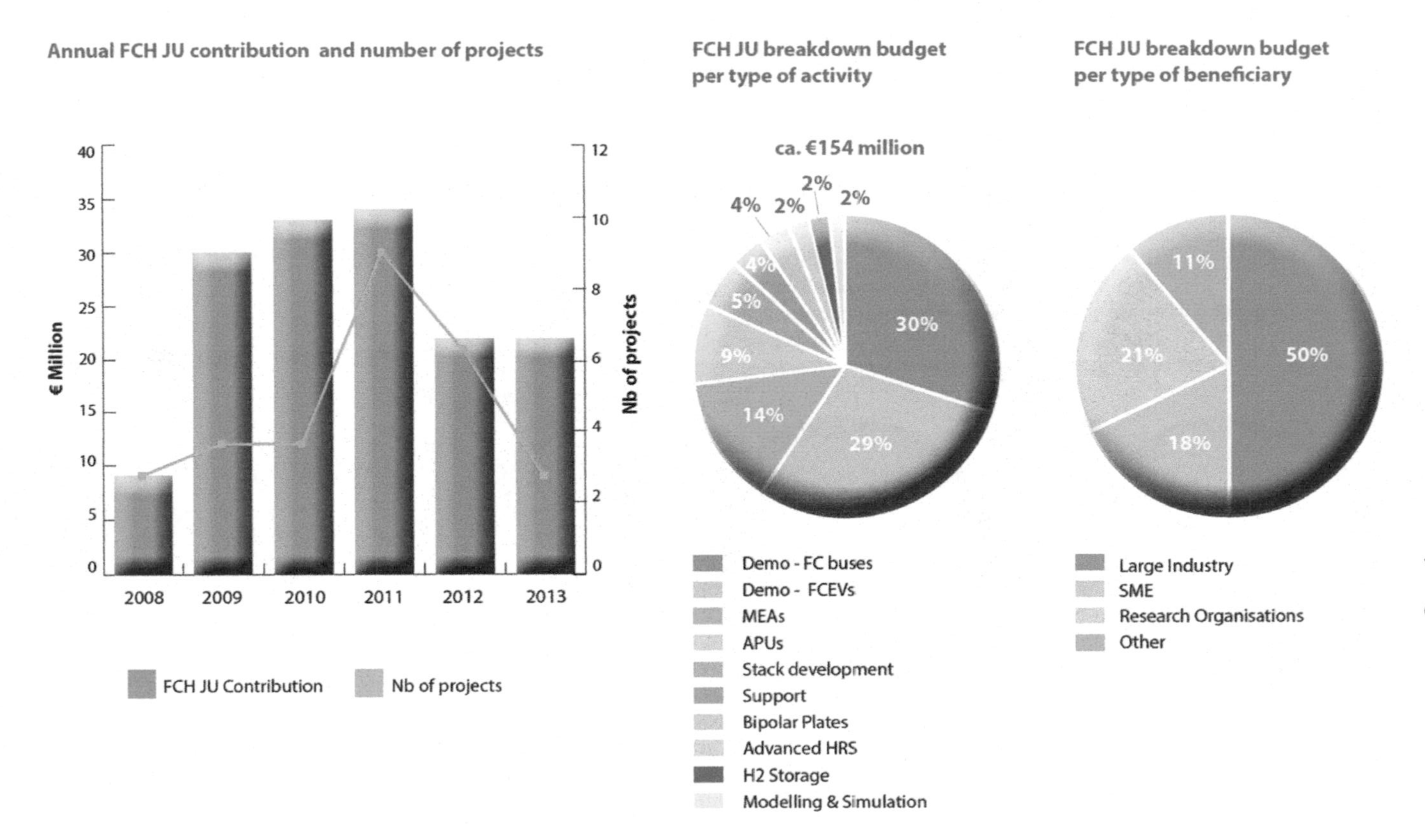

Figure 9.3 The FCH JU projects on transport and refueling infrastructure.

Source: FCH JU.

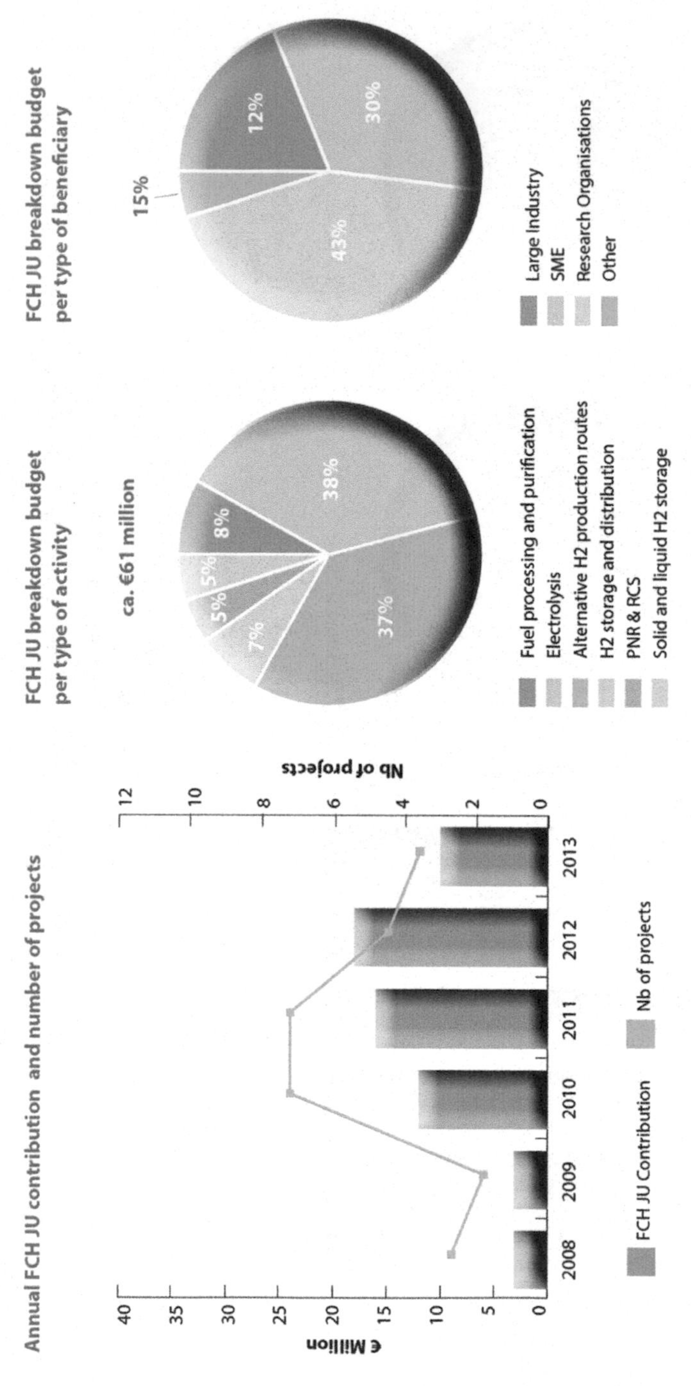

Figure 9.4 The FCH JU projects on hydrogen production and distribution.

Source: FCH JU.

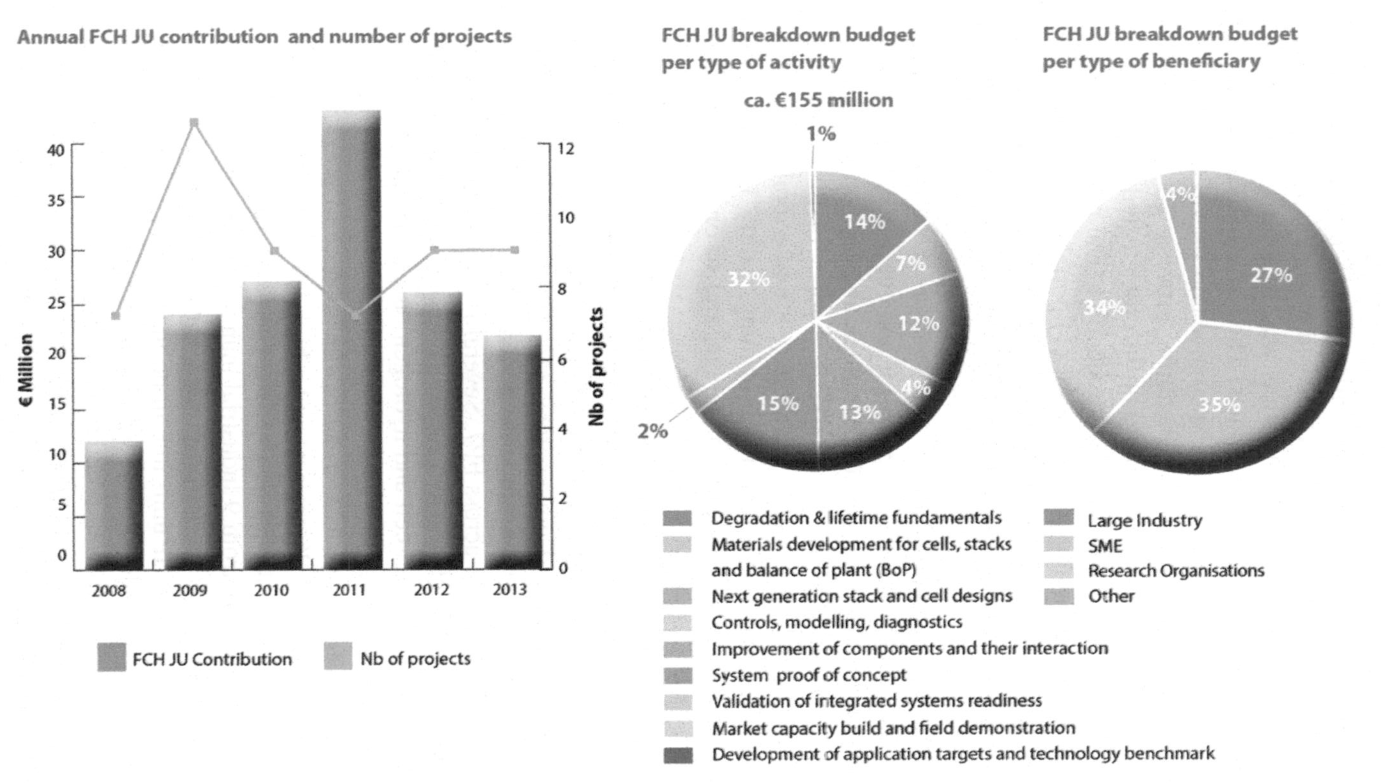

Figure 9.5 The FCH JU projects on application for stationary power generation and CHP.

Source: FCH JU.

In order to build up and sustain the early manufacturing and supply base for fuel cell technology, the FCH JU created a strategic pillar entitled early market. The budget of this application was focused on short-term demonstrations and ready-to-market products. The early market pillar was an important tool for gaining operating experience and providing feedback into the technical development and manufacturing process. The aim for the projects' demonstrations was volume build-up in order to reduce the costs and pave the way for commercial market introduction. The pillar wanted to be an umbrella for the technology readiness of (1) portable and micro fuel cell for diverse applications, (2) portable generators or back-up power, and (3) handling vehicles and associate technology. More details are given in the Figure 9.6.

The goals of the cross-cutting pillar was to evaluate the socio-economic, environmental and energy impact of hydrogen and fuel cell technologies, monitor the research and development programme and support the growth of the industry, especially SMEs. The TCH JU budget of this pillar was around 19 M Euro, which means approximately 6–8% of overall budget for 18 projects focusing mainly on education, training, safety issues, socio-economic aspects, hydrogen underground storage, and so on at the level of 2013, Figure 9.7.

Established in 2008, the FCH JU has supported 169 projects to date, which are starting delivering very important results. More than 500 organizations have been participating in the activities of the FCH JU (among these 167 research centers and universities, 342 private enterprises). The FCH JU has also been successful in attracting SMEs, which accounts for 27% participation compared to 18% across FP7. Several FCH JU funded projects highlight the impact of fuel cell and hydrogen technologies. For example, the *CHIC* project integrates 26 fuel cell and hydrogen buses in public transport in five locations across Europe and has already demonstrated a significant reduction in fuel consumption of over 50% with respect to previous bus generations demonstrated in FP6 projects and a high level of availability of the 13 hydrogen refuelling stations (>98%). The *ene.field* project will install about 1000 micro-CHP units from 9 industrial manufacturers for residential use in 12 Member States. *SOFT-PACT* project, through the leader E.ON, intends to deploy 100 micro-CHP units in Europe and therefore, demonstrates an electrical efficiency of at least 60%, which can provide the EU citizens (real end users) with an efficient product that can reduce the household expenditures regarding energy bill. *FITUP* project is a demonstration project in which a total of 19 market-ready fuel cell systems from two different

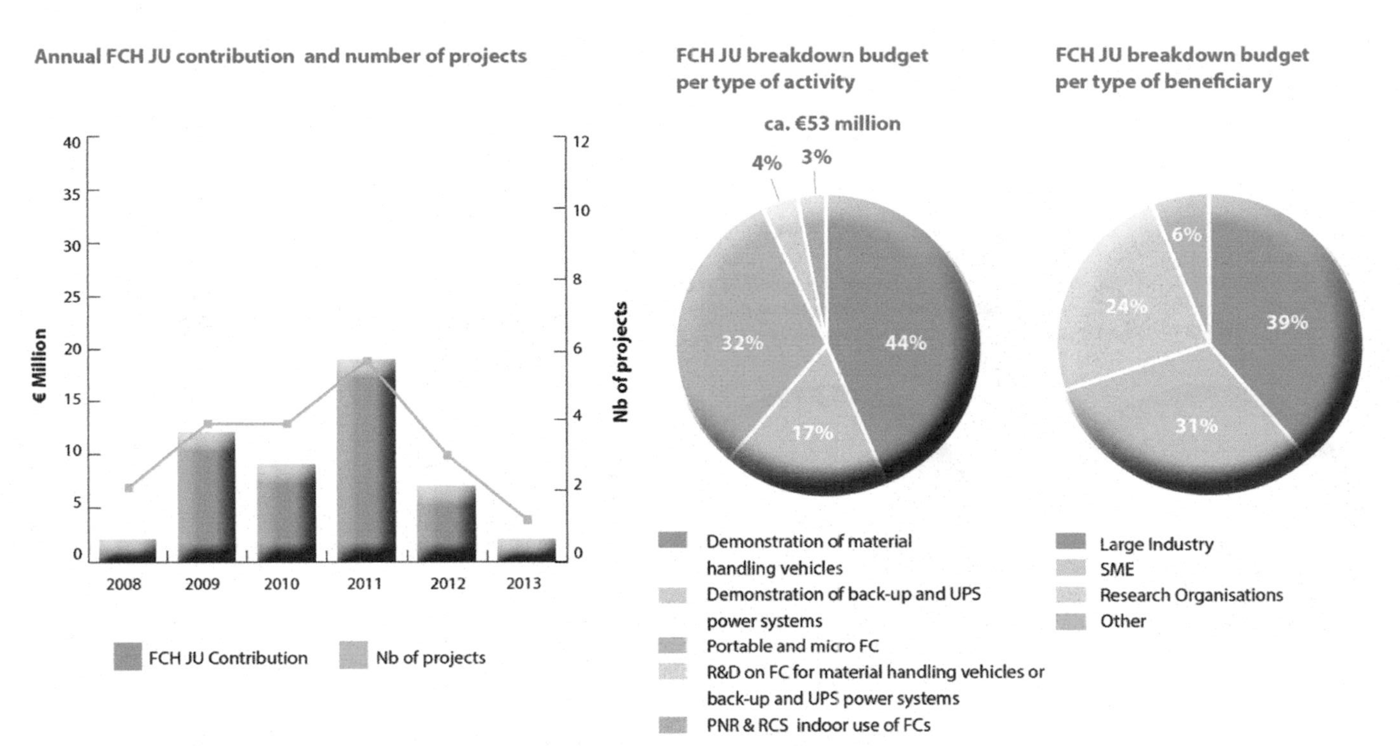

Figure 9.6 The FCH JU projects in early markets pillar.

Source: FCH JU.

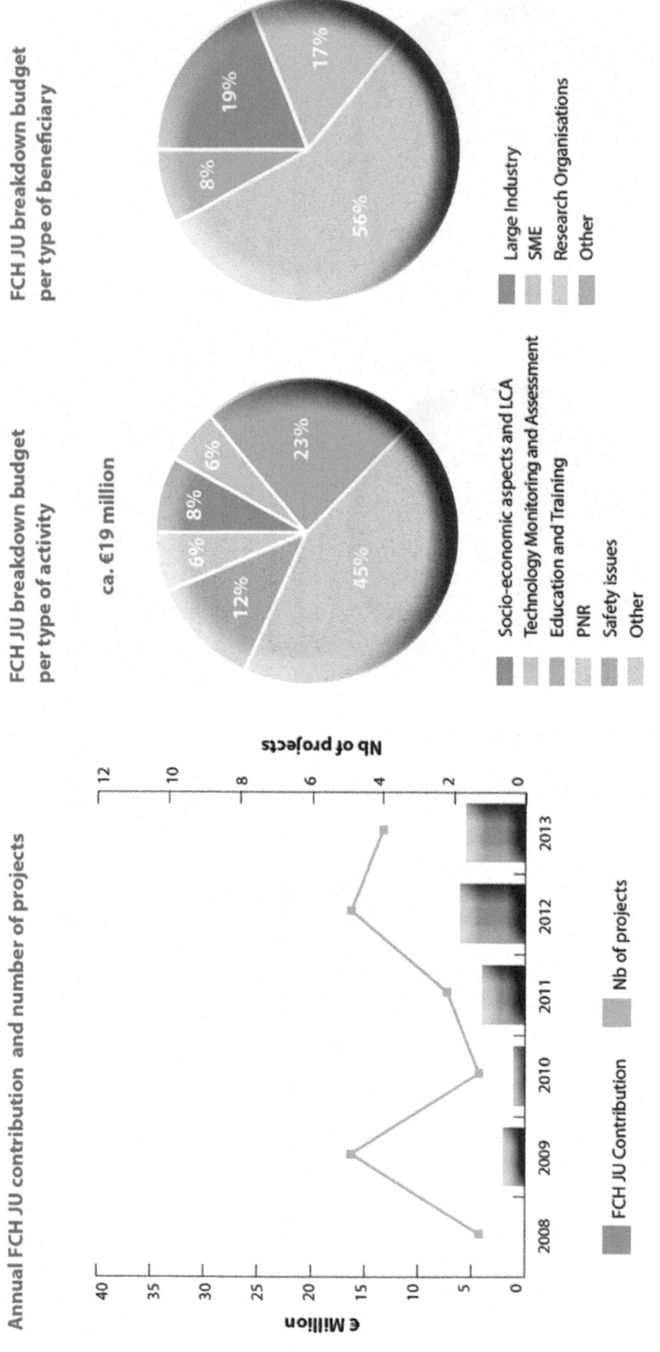

Figure 9.7 The FCH JU projects in cross-cutting.

Source: FCH JU.

suppliers are installed as backup power sources by final. This technology shows advantages compared to incumbent technologies (diesel generators and batteries systems) allowing for less noisy, cleaner, and longer duration operation. There is still a lot to do but huge progresses have been achieved notably as regards improving materials' performance and durability, as well as reducing costs of components and systems in both transport and energy applications [7].

Various hydrogen and fuel cell technologies received support during FP5 (1999–2002): Polymer Electrolyte Membrane (PEMFC), including Direct Methanol (DMFC), Solid Oxide Fuel Cell (SOFC), and Molten Carbonate Fuel Cell (MCFC). The hydrogen research funding was granted for production technologies for both conventional and renewable resources, storage technologies, and issues surrounding the introduction of hydrogen as an energy carrier. Demonstration projects for both stationary and transport applications have been funded in the framework program [8]. The authors accounted a number of 66 projects.

For hydrogen, several strategic topics for research was pursued in FP6 (2002–2006): clean production, storage, basic materials, safety (pre-normative regulations), and preparing the transition to a hydrogen energy economy. In the area of fuel cell systems, the funds were focused on the cost reducing and improving of the performance, durability, and safety of fuel cell systems for stationary and transport applications, to enable them to compete with conventional combustion technologies. There was included materials and process development, optimization and simplification of fuel cell components and sub-systems as well as modeling, testing, and characterization protocols. The researchers also included validation and demonstration activities to gain experience that was introduced again into technology development and deployment, as well as providing first-hand training to stakeholders and end users [9]. In this research and development framework program, about 84 projects were counted.

9.4 The Fuel Cells and Hydrogen 2 Joint Undertaking

The Fuel Cells and Hydrogen 2 Joint Undertaking is the successor to the Joint Undertaking in the field of Fuel Cells and Hydrogen launched under the Seventh Framework Programme (FP7). It will continue to sustain the research, technological development, and demonstration activities for the time frame 2014–2020, Council Regulation (EU) No 559/2014 of 6 May 2014. According with the explanatory memorandum that accompanied the Council

Regulation, the general objective of the FCH 2 JU for the period of 2014–2024 is to develop strong, sustainable, and globally competitive fuel cells and hydrogen sector in the EU, in particular to: (1) reduce the production cost of fuel cell systems to be used in transport applications, while increasing their lifetime to levels competitive with conventional technologies, (2) increase the electrical efficiency and the durability of the different fuel cells used for power production, while reducing costs, to levels competitive with conventional technologies, (3) increase the energy efficiency of production of hydrogen from water electrolysis while reducing capital costs so that the combination of the hydrogen and the fuel cell system is competitive with the alternatives available in the marketplace, and (4) demonstrate on a large scale the feasibility of using hydrogen to support the integration of renewable energy sources into the energy systems, including through its use as a competitive energy storage medium for electricity produced from renewable energy sources [10].

The scope and the objectives of the research, development, and demonstration activities are outlined in the framework document named *Multi Annual Work Plan* (MAWP), and at least one call will be done per year based on the *Annual Work Plan* (AWP). The programme has been designed and carried out by and in cooperation with the stakeholders: industry, including SMEs, research centers, and universities, together with the Member States and Associated Countries, and regions and municipalities in Europe. The structure of the MAWP is shown in Figure 9.8, illustrating the breadth of activities expected across the two innovation pillars: transportation and energy. The transport include the refueling infrastructure activities. The energy pillar encompasses hydrogen production and distribution, stationary power generation and CHP. Both of them include early market activities and are supported by the crosscutting activities, including market support measures, public awareness and education, and various assessments.

The objective of the transport pillar is the acceleration of the commercialization of hydrogen and fuel cell technologies for use in a range of transport applications. In concordance with SET-Plan, these technologies will play a critical role in the decarbonization of Europe's transport sector. Renewable hydrogen and fuel cells will provide the means to achieve a greenhouse gas zero-emissions transport sector. The efforts will be focused on reducing the production costs of fuel cell systems and increasing their lifetime in order to be competitive with conventional incumbent technologies. The objective of the energy pillar is to accelerate the commercialization for the stationary applications, hydrogen production, and storage and distribution technologies [11].

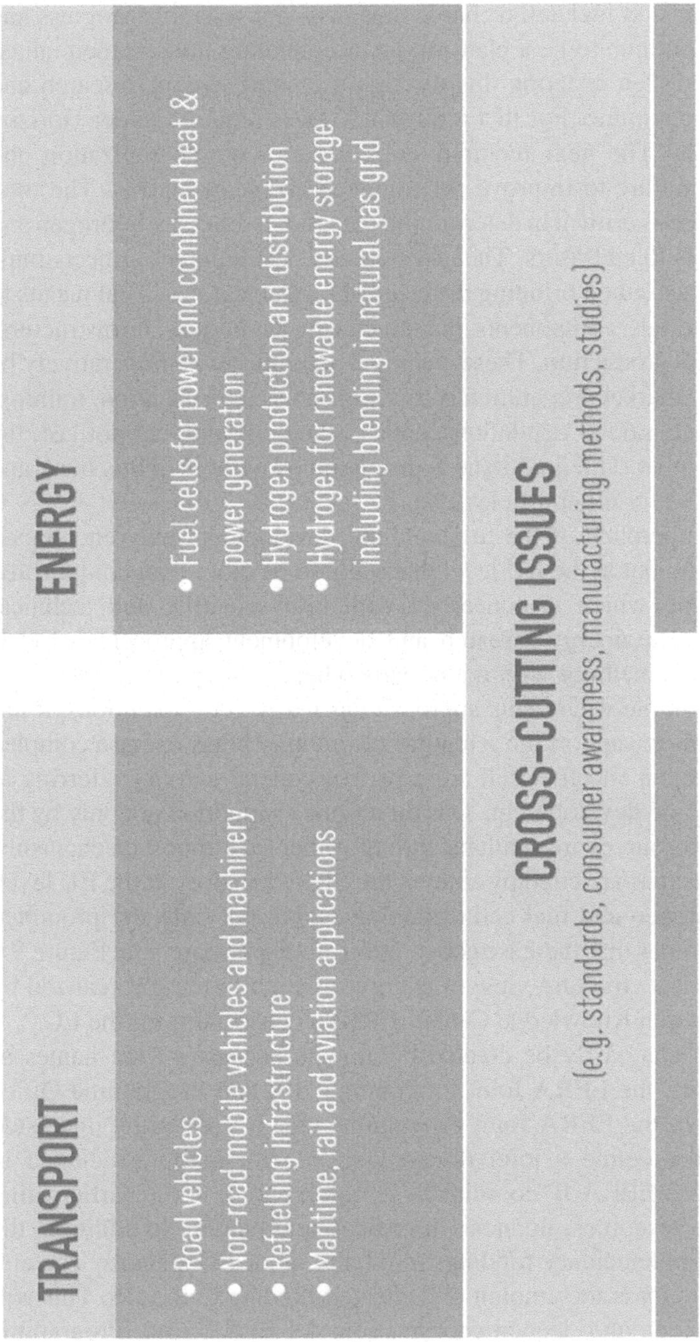

TRANSPORT

- Road vehicles
- Non-road mobile vehicles and machinery
- Refuelling infrastructure
- Maritime, rail and aviation applications

ENERGY

- Fuel cells for power and combined heat & power generation
- Hydrogen production and distribution
- Hydrogen for renewable energy storage including blending in natural gas grid

CROSS-CUTTING ISSUES

(e.g. standards, consumer awareness, manufacturing methods, studies)

Figure 9.8 The FCH JU activities under the Multi Annual Work Plan.

Source: FCH JU.

The hydrogen and fuel cell technologies made considerable progress and FCH JU must continue to be a platform for accelerating the commercialization of them and then enacting this through a comprehensive research and development programme. For that a further work is required under Horizon 2020 programme. The next required steps for pre-commercialization and commercialization are to improve performance and reduce costs. The total cost of ownerships is critical in determining the attractiveness of hydrogen and fuel cell products to end-users. The research and development projects must continue to be focused on bringing the costs of products down. That means to improve the materials, components, manufacturing techniques, infrastructure, and also the mass production. These technical activities must imperatively be accompanied by market preparation activities such as the education, training, public acceptance, safety, regulations, codes, standards and, but not last, the vulnerability combat. The final issue is discussed by authors in this book and will be the subject for other articles also.

The simple approach of the research and development may encompass some basic issues, but at the EU level that seem to be more than complicated for an researcher, which is concerned with him scientific and technical preoccupations. The energy research and development approach in EU is characterized as something with many parts where those parts interact with each other in multiple ways. This approach has really good intentions, it has really useful tools to support the scientific community but remains a complex issue. The hydrogen and fuel cell are a part of general activity referring to energy research and development. This direction is supported not only by the Horizon 2020. Resources are available through other EU support mechanisms, in the first paragraphs are cited by authors the SET-Plan also. At the EU level, research of hydrogen and fuel cell technologies, but not only, are promoted through programmes that increase cooperation among researchers. Figure 9.9 is an organizational structure view of energy research in the UE realized by the Energy Research Knowledge Center (ERKC) consortium for the EC.

A confusion can easily be created by anyone because of the names of the two programs, the EERA Joint Programme (JP) and Programme Office of FCH JU. First, the EERA Joint Programme (JP) is created by interested organizations that define a joint research agenda for a topic included in the SET-Plan. The EERA JP coordinates research based on the participating institutions own resources, financial, logistical, and human. In addition, the JP can obtain supplementary funding from national or EU sources. The aim is to gradually increase the amount of dedicated funding to the JPs. This will allow a JP to widen and deepen coordination. An EERA Joint Programme

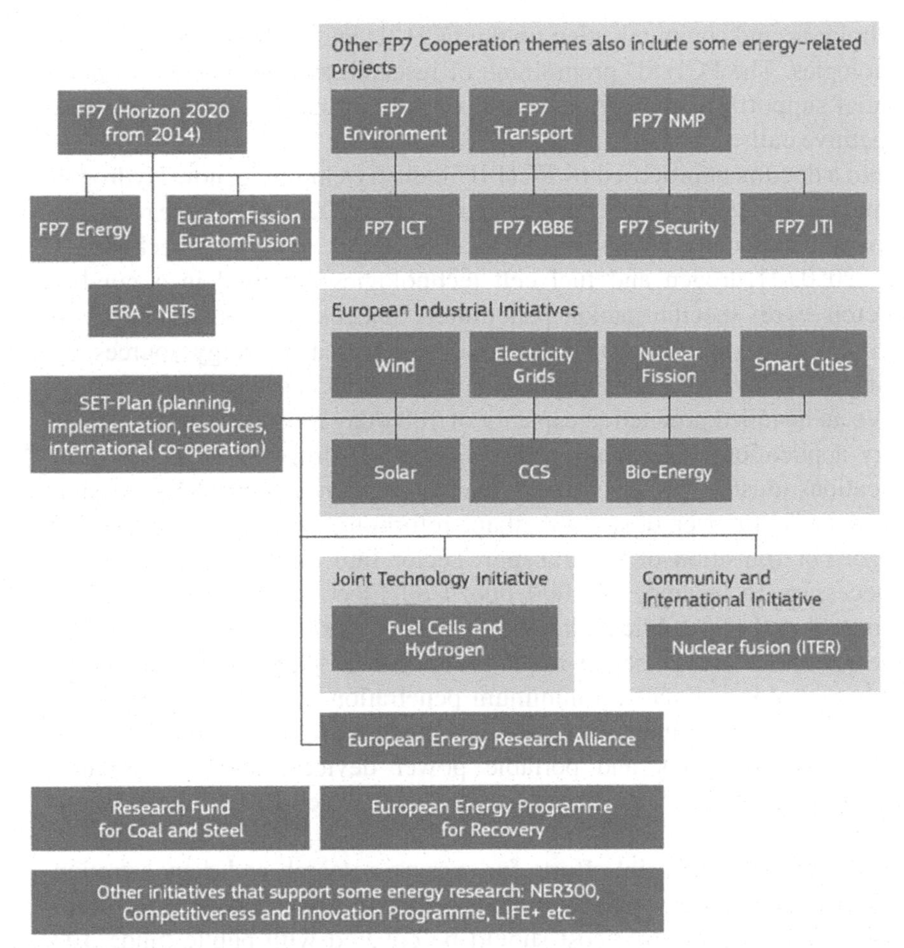

Figure 9.9 The complex organizational structure of energy research and development in EU.

Source: ERKC.

consists of: the programme steering committee, the programme management board, and the sub-programmes divided into research areas. A participating institution can be either a full or an associate participant of a JP. Second, the Programme Office is in charge of the daily management of the FCH JU and executes all its activities, from project management to financial matters and communication, under the responsibility of the Executive Director. The key responsibilities of the Programme Office are: organization of the calls for proposals, selection of projects, and management of funding; management of the agenda of the FCH JU in coordination with members and other stakeholders; and communication on the FCH JU and fuel cell and hydrogen technologies. The FCH JU programme of research and innovation provides financial support in the form of grants to participants following open and competitive calls.

Into a document proposed by FCH JU Industry Grouping, now Hydrogen Europe, was presented the Technology Roadmap to 2020 [12]. These roadmap put forward concrete action plans aimed at raising the maturity of technology up to 2050. Hydrogen and fuel cell technologies are used in a number of sectors, with specific market penetration objectives by 2020. One is to contribute to the integration of intermittent renewable energy sources by proving 500 MWe cumulative hydrogen conversion capacity. Other item is to have an installed production capacity of 100 t/day renewable hydrogen for energy applications. A proportion of 50% from hydrogen used into energy applications must be produced from renewables or near zero-CO_2-emission sources, nuclear power, or steam methane reforming with carbon capture and storage. For transition of the transport sector into electric mobility, there are necessary to incubate 500,000 Fuel Cell Electrical Vehicles (FCEVs) and more than 1,000 public hydrogen refuelling stations. The transformation of the European energy sector requires more than 50,000 fuel cell micro-CHP householders. Finally, a minimum penetration in the early market for applications such as material handling vehicles (20,000 units), beak-up/ UPS (20,000 systems), and portable power devices (250,000 systems) can contribute to the demonstration that the fuel cell-based solutions are cost efficient.

The indicative financial efforts for support research and demonstration activities or programmes was estimated to about 6.4 bn. Euro for the period 2014–2020. This indicative cost should be covered with public funds, EU and Member States, and from private investments. Second, a dedicated approach to market introduction effort using innovative financing schemes will be necessary to cover around 11.5 bn. Euro. The total financial joint public private efforts and investments in the development and deployment of hydrogen and fuel cell technologies were estimated at around 17.9 bn. Euro for the aforementioned period. From this estimated amount total EC contribution of 2.5 to 4 bn. Euro is need, together with Member States contributions of 2 to 4 bn. Euro and an investment of 10–14 bn. Euro from the private sector.

Why such efforts are suggested? The European ambition for a low carbon economy means sustainable, secure, and affordable energy supply and service. There is need for a joint effort where the clean and innovative technologies are essential for a successful transition. The hydrogen is an integrant and important part of the European solution and not an item of the problem. The following decalogue are lined up the situation of the hydrogen and fuel cell technology in Europe, seen through the eyes of the industry stakeholders (New Energy World Industry Grouping [NEW-IG] was the name at the moment when thy prepared the document, now it is Hydrogen Europe):

1. Fuel cell and hydrogen technology is vital for the future European economy.
2. Decisive action is needed to maintain Europe's global technology leadership for the future.
3. A purely market-driven approach alone will not enable the introduction of clean technology.
4. Public private partnership is the appropriate structure to support the technological shift.
5. Joint public and private effort need for hydrogen and fuel cell technology breakthrough across sectors reach 17.9 bn. Euro for 2014–2020.
6. Investment focus is twofold, improving the competitiveness of hydrogen and fuel cell technology solutions, and increase the share of renewable sources in the hydrogen production mixture.
7. Combined public and private investment is needed for all stages of the innovation cycle, from research and development to first-of-a-kind commercial references.
8. Bringing clean technologies to the point of market breakthrough might require a shift from technology to sector support.
9. New financial investments are needed to finance first-of-a-kind commercial applications and support market introduction.
10. Hydrogen and fuel cell technologies should benefit from various European programmes.

Into a document published recently by International Energy Agency (IEA), there are mentioned similar key actions for the next 10 years in order to stimulate the development and implementation of hydrogen and fuel cell technology [13]:

- Encourage fuel efficiency and low greenhouse gases emission technologies across all energy sectors using a stable policy and regulatory framework that is important for raising market certainty for entrepreneurs and investors.
- Stimulate investment and early market deployment of hydrogen and fuel cell technologies and infrastructure in order to bring down the costs and overcome the market barriers.

- International codes and standards harmonization are necessary for safe and reliable handling and metering of hydrogen in end-use applications.
- Keep up supporting technology progress and innovation by unlocking public and private funds for research, development, and demonstration.
- Improve understanding of regionally specific interactions between different energy sectors through integrate modeling approaches to quantify benefits of energy system integration.
- Accelerate activities directed at the developing the capture and storage of CO_2 from fossil-driven hydrogen production into mature business activities.
- Putting first tens of thousands of fuel cell electrical vehicles on the road along with hydrogen production and refueling infrastructure, 500 to 1,000 refueling stations is recommended.
- Development of financial instruments and innovative business models that de-risk the hydrogen infrastructure developed for fuel cell electrical vehicles market introduction.
- Establish regulatory frameworks that remove barriers to grid access for electricity storage systems including power-to-gas (fuel) applications. Where is regionally relevant, establish a regulatory framework for the blending of hydrogen into the natural gas grid.
- Increase data on resources available and costs for hydrogen generation at national and local levels. Analyze of the potential future available of curtailed electricity as a function of intermittent renewable energy integration, others power system options, and competing demands for any surplus renewable energy.
- Address potential market barriers where opportunities exist for the use of low-carbon hydrogen industry, for example, in refineries or ammonia plants.
- Education programs and intensive information campaigns to increase awareness-raising.

Regarding the current performances of the hydrogen conversion, transport, distribution, and storage technologies, IEA reviewed more papers and presented some data, represented in Table 9.1.

Aforementioned are completed with other data presented by FCH JU in a study on the development of water electrolysis in the EU, Table 9.2 [14]. In the list of electrolyzer suppliers can be observed various technologies for water electrolysis, alkaline, anion and cation exchange membranes, capacities from some cubic metre per hour till at some hundreds, and output pressures from atmospheric conditions till at 31 bars (g).

Table 9.1 Current performances of the hydrogen technologies

Technology	Power	Efficiency	Investment	Live Time	Stage
Alkaline fuel cell	Max. 250 kW	50% (HHV)	200–700 USD/kW	5,000–8,000 h	Early market
PEM fuel cell (stationary)	0.4–400 kW	32%–49% (HHV)	3,000–4,000 USD/kW	Approx. 60,000 h	Early market
PEM fuel cell (mobility)	80–100 kW	Max. 60% (HHV)	Approx. 500 USD/kW	Max. 5,000 h	Early market
SOFC	Max. 200 kW	50%–70% (HHV)	3,000–4,000 USD/kW	Max. 90,000 h	Demon-stration
PAFC	Max. 11 kW	30%–40% (HHV)	4,000–5,000 USD/kW	30,000–60,000 h	Mature
MCFC	kW–MW	>60% (HHV)	4,000–6,000 USD/kW	20,000–30,000 h	Early market
Compressors-18 MPa	–	88%–95%	Approx. 70 USD/kWH$_2$	20 years	Mature
Compressors-70 MPa	–	80%–91%	200–400 USD/kWH$_2$	20 years	Early market
H_2 liquefier	15–80 MW	Approx. 70%	900–2,000 USD/kWH$_2$	30 years	Mature
H_2 on-board storage tank (70 MPa)	5–6 kg H_2	–	17–33 USD/kWh	15 years	Early market
H_2 pressurised tank	0.1–10 MWh	–	6,000–10,000 UDS/MWh	20 years	Mature
H_2 liquid storage	0.1–100 GWh	Boil-off stream: 0.3% loss per day,	800–10,000 UDS/MWh	20 years	Mature
H_2 pipeline	–	95% (incl. comp.)	300,000–1,500,000 USD/km (depend on diameter)	40 years	Mature

HHV–Higher heating value, MCFC–Molten carbonate fuel cell,
PAFC–Phosphoric acid fuel cell, PEM–Proton exchange membrane, SOFC–Solid oxide fuel cell.
Source: IEA.

Table 9.2 List of electrolyzer suppliers

Company (Country)	Type	Product	Capacity (Nm^3/h)	Output Pres. (bar_g)	Purity H_2	Consumption (kWh/kg)	Efficiency (%)
Acta (IT)	AEM	*EL1000*	1	29	99,94	53,2	63
AREVA (FR)	PEM	–	20	35	99,9995	55,6	60
CETH2 (FR)	PEM	*E60 cluster*	240	14	99,9	54,5	61
ELT Electrolyze (DE)	Alc.	–	330	Atm.	99,85	51	65
Erredue (IT)	Alc.	*G265*	170	30	99,5	59,5	56
H2 Nitidor (IT)	Alc.	*200 Nm3/h*	200	30	99,9	52,3	64
H-TEC SYSTEMS (DE)	PEM	*EL30/144*	3,6	29	–	55,6	60
Hydrogenics (BE, CA)	Alc.	*HyStart60*	60	10	99,998	57,8	58
Idroenergy (IT)	Alc.	*Model120*	80	5	99,5	52,4	64
IHT Industrie Haute (CH)	Alc.	–	760	31	–	51,2	65
ITM Power (UK)	PEM	*Hpac40*	2,4	15	99,99	53,4	62
NEL Hydrogen (NO)	Alc.	–	485	Atm.	>99,8	50	67
McPhy (DE)	Alc.	*60 Nm3/h container*	60	10	99,3	57,8	58
Proton OnSite (US)	PEM	*Hogen C30*	30	30	99,9998	64,5	52
Siemens (DE)	PEM	*SILYZER200*	~250	–	–	60	55
Teledyne Energy Systems (US)	Alc.	*SLM 1000*	56	10	99,9998	–	–
Wasserelektrolyse Hydrotechnik (DE)	Alc.	*EV150*	225	Atm.	99,9	58,7	57

AEM–Anion exchange membrane
Alc–Alkaline
PEM–Proton exchange membrane
Source: FCH JU.

The FCH JU and hydrogen and fuel cell key stakeholders indicated that there are five critical challenges that need to be overcome in order to be successful: the commercialization rate, infrastructure, the continuation and maturity of research, competition with other regions and technologies, and public acceptance [15].

The expected date of commercialization has systematically fallen behind promises. Missing a credible and accurate time path is a risk in attracting and retaining private investors and well-intentioned public authorities.

The fuel cell and hydrogen fleets depend fully on a widespread fueling infrastructure to attract customers. At this moment the fuel cell and hydrogen mobility is very fragmented. The energy and fuel companies will invest only if there is a sizeable market and the car manufacturers will produce vehicles at scale only if the necessary infrastructure is present. The problems can be solved only by cohesive and coalition-led activities. In the opinion of ones, the German H_2 Mobility is an example, but is too early to claim the success.

The hydrogen and fuel cell research and development is highly dependent by regional, national, and European funds and grants. Research and development is vital for the early market implementation but the research targets and quality of the results do not always correspond with the priorities of companies implied in the commercialization of this technologies. This limits the impact of the activities done and capacity to absorb private/industrial funds.

At list at the level of year 2015, the European progress is behind of other hydrogen and fuel cell development hubs Japan-South Korea and US-Canada, where the early market introduction is seen to be more productive. The Europeans have an advance only in fuel cell buses sector. The industry sector should be careful that the nucleus of knowledge development does not shift out of Europe.

The hydrogen and fuel cell technologies and advantages are unfamiliar to people and also for many potential customers. These technologies have not received widespread public attention, despite of it very focused on the environment attention.

9.5 Hydrogen in Europe and the Rest of the World

The preoccupation for the hydrogen and fuel cell technologies are in all world not only in Europe. The authors have located three main hubs regarding the research, development, demonstration, or introduction on the (early) market: Japan-South Korea, US-Canada, and EU, without claiming to be universal and generally valid, this group was made taking into account the visibility and constant concern. Of course, we do never neglect the efforts made in South Africa, Australia, South America, China, or Russia. The European constant and coherent support for these technologies through the FCH JU is very useful but is enough? In the next paragraphs will be mentioned some progress and facts existing in the world.

In one recent review of the *E4tech* are presented some realities regarding fuel cell technologies [16]. In 2015 were shipped 71,500 fuel cell units, which cumulated 342.6 MW. The shipments, both in number and power, growth constantly from 2008 until 2014: 9,500 (51.1 MW), 14,400 (86.5 MW), 17,700 (91.2 MW), 24,600 (109.6 MW), 45,700 (166.7 MW), 66,800 (215.3 MW), and 63,800 (185.4 MW). These fuel cells are used specially for stationary applications (62% and 69% in 2014 and 2015) followed by portable and transport applications. The shipment by fuel cell type is dominated by PEMFC with a share that increased from around 60% in 2009–2010 to 90% in the past years. The shipments by region are dominated by Asia, which covered more than 60% of demand in the previous years. If until 2012, Europe was on the second place as the number of fuel cell units used in various applications, in the previous years this position has been lost in favor of the North America, which was all time in the second place when taken into account the installed power.

The authors of aforementioned review remarks that Europe remains a slight anomaly. With strong companies, with national supports, especially in Germany, with public-private funds available from FCH JU, the European development is nevertheless much lower than other regions, only around 10% of shipments in both unit number and installed capacity. This reflects an unhealthy commercial subsidy regimes. The uncertainty in European energy markets, turmoil in the energy utilities or the austerity regimes contribute in some proportions contribute to this landscape. The exception is only the fuel cell application in bus mobility. At this moment, more fuel cell buses are operating in Europe than anywhere else. But also this pole position is targeted by China's intentions to have more fuel cell buses on the roads.

In the United States, the main activities for the research, development, and demonstration of hydrogen and fuel cell technologies is conducted under *US DOE Hydrogen and Fuel Cells Program*. The Program works in partnership with industry, academia, national laboratories, federal and international agencies. The DOE supports the research and development of hydrogen and fuel cell technologies as part of a portfolio of renewable and energy-efficient technologies. The Office of Energy Efficiency and Renewable Energy has spent nearly 1.5 bn. USD in the past decade on fuel cell and hydrogen research, development, and demonstration. This investment has led to more than 500 patents and about 40 commercial technologies being introduced on the market and 65 emerging technologies that are anticipated to be in the market within next years. The Energy Policy Act of 1992 was first national legislation in the US that called for large-scale hydrogen research and development. Between 1990 and 2002, this specific program have cumulated in fact two separate programs for hydrogen and fuel cell. Starting with 2003, the aforementioned office combined hydrogen and fuel cell programs into one. In 2004, the Hydrogen *Fuel Initiative (HIF)* began and continued for five years. Starting with 2009 the hydrogen and fuel cell research, development, and demonstration activities continue in the four DOE offices, as coordinated efforts. *The Solid State Energy Conversion Alliance (SECA)*, within Fossil Energy Office, focused on MW scale solid fuel cell development and was not part of the HIF. The Solid State Energy Conversion Alliance (SECA) has been in collaboration between the US Federal Government, private industry, academic institutions, and national laboratories devoted to the development of low-cost, modular, and fuel-flexible solid oxide fuel cell technology since1999. Overall, the funding for hydrogen and fuel cell program is almost 2.5 bn. USD while SECA has provided over 460 million USD, seem to be a big amount but is less than 0.9% of the total Department of Energy (DOE) budget [17]. In 2013, to help address the challenge of hydrogen infrastructure, DOE co-launched *H2USA*, a public-private partnership focused on the widespread commercial adoption of FCEVs. H2USA currently consists of 45 participants, including the state of California, as well as developers, car companies, and hydrogen providers.

Steven Stoft and César Dopazo remarked that the US centralized research approach has been much more successful than the less-structured and fragmented EU pursuit. The running programs in the US could be compared with that of an orchestra, with well-trained musicians (15 national laboratories, nearly 300 university research groups and powerful companies) actively coordinated by the DOE, the coordination at the European member state level is hampered by many orchestras with many conductors, which insist on playing their own scores, with inefficient spending of funds and not profiting from synergies. The EC Joint Research Center activities in support of EC policy measures are not comparable to those of the US national laboratories [18].

For over 30 years, Japan has accumulated experience and accomplishments in the hydrogen and fuel cell energy research and development. The development and demonstration of hydrogen and fuel cell technologies is conducted under organization *New Energy Development Organization (NEDO)*, established in 1980. NEDO coordinates and integrates a wide range of the technological capabilities from private enterprises to academia or research facilities and organizes technology development activities as national projects to realize fundamental technologies (including technology demonstrations). NEDO National Project Activities undertakes the development of new energy, here it is included fuel cell and hydrogen. The NEDO actions contributing to the building a new social system where hydrogen is not a simply fraction as an alternative to existing energy. Looking at the medium and long term, Japanese are expected big things from hydrogen as a form of clear energy.

Due to the realities of Japan: 141,000 micro-CHP installed; 700 Toyota Mirai FCEVs on the roads with a prognosis of 2,000 in 2016, other 3,000 in 2017, and a total of 30,000 in 2020; a secured budget for a total of 81 new HRSs and 8 Euro/kg the price of H_2 at the pump: the year 2015 was called "first year of hydrogen society". NEDO vision has three steps to deeply understanding in hydrogen and fuel cell development toward realizing the hydrogen society [19].

First is to make hydrogen and fuel cell familiar and contributing to the "final push" for commercialization. Japan's *Ene-Farm* program is arguably the most successful fuel cell commercialization framework actions in the world. Hydrogen made from natural gas have been utilized for electricity supplying and hot water using household fuel cell systems. The program supported the deployment of residential micro-CHP units and is providing proof that long-term public-private partnerships can push new technology into the marketplace. Japan's interest in residential fuel cell applications dates to the 1999. The *Millennium Project* included support for materials and fuel cell research. The national targets are 1.4 million residential units by 2020 and 5.3 million by 2030 (about 10% of Japan's homes). A large-scale private-public demonstration approach, totaling 3,300 units, began in 2005 and was successful enough to justify commercial launch in 2009 [20].

The other step is the hydrogen infrastructure development for FCEVs. The Japanese have called the FCHV as "ultimate eco-car" because this vehicle does not exhaust any other than water. With the aim of making the technology more widespread, they are working on the development of low-cost equipment. NEDO is working to develop materials used in fuel cell vehicles and methods to evaluate such materials. Technology for establishing a hydrogen infrastructure is also being developed and regulations are being reviewed.

The next step following the residential systems and mobility, NEDO is focused on hydrogen power generation. A significant expansion of hydrogen utilization is necessary through a large-scale power generation. The hydrogen utilization into gas turbine power generation can significantly expand it usage and reduce CO_2 emissions. NEDO has embarked on technological development to cover unused energy from overseas into hydrogen and transport it long-distance to Japan. The challenge is the hydrogen long-distance transport and storage because its energy density per volume is about one-third that of natural gas. NEDO reviews a variety of ways to transport and store of the hydrogen but the most promising are the shipment as liquid gas and as organic hydrides.

NEDO's newest projects on technology development for the realization of hydrogen society are about (1) large-scale hydrogen systems, 2015–2020, and (2) renewable energy-driven hydrogen systems, 2014–2017. The first projects aims are to establish the flexibility of the energy supply systems and contribute to ensuring energy security. The project demonstrations will include: the liquefied hydrogen production, storage and marine transport from brown coal, the hydrogen production from unused or underutilized resources and transport it to consumers using organic chemical hydride, a 1 MW-class gas turbine that uses hydrogen as fuel with the aim of efficient use of electricity and heat at the local level, and research and development of a 100 MW-class hydrogen and natural gas co-firing turbine and power plant. The second project will establish the model, an implementable model for a "Power-to-Gas" energy system. The project will put together the intermittent renewable energy systems and hydrogen storage solution.

9.6 Final Remarks

As it has already been mentioned in the first part of this chapter, hydrogen is a decisive issue, a vital matter, with economic, geopolitical, and societal ramifications. The main roles of these pages are to outline the institutional efforts of public actors regarding the hydrogen research and development. For many this topic seems to cover the energetic and technical aspects, but it is a false impression, the hydrogen ramifications are more complicated. The institutional supports without exposing to the public will not work, in isolation any scientific or technical approach is unlikely to germinate. The main weakness of the public programs for hydrogen and fuel cell supporting is that those start from the middle of approach, the real people needs are omitted in the favor of technical and market aspects.

Most people were not familiar with hydrogen and fuel cell, they have heard of hydrogen and have some associations. When is discussed the possibility to use hydrogen as a fuel, it is associated with the risk of explosion. There is a severe lack of understanding the whole chain: from the need to increase use of renewables, to the problem of intermittent energy supply by renewables, to the need to store energy, and to the role of hydrogen in this low carbon equation. After people will have information on this topic, they will be mostly enthusiastic [21].

Because at continental scale (EU) it is needed to avoid geopolitical and societal fragmentation, there is needed to develop of a series of actions for the concatenation of non-active stakeholders with the potential for hydrogen community. The FCH2 JU budget is 50–50 (public-industry), but this public money are from all Europe Union, Members States, and Associated Countries, for medium and long terms way somebody believe that states and people will want to sustain financial projects where they are absent? What are their benefits?

Industry is a very important partner and must be sustained in order to be competitive and to create affordable products. The industry is one heavy element, but the economy of hydrogen is more complex and here the industry is only one gearwheel of gearbox. We will not have success if we take care only for industry and we forget many other elements necessary in the real hydrogen economy. In this hydrogen economy industry is not alone, all actor are important and is like a clock, if there missing a wheel will not work forever, in the supply-and-demand equation the real people needs are the heart of the system.

In order to be in the pole position and to decrease the vulnerability, the EU must concomitantly approach three issues: (1) to build-up an energy landscape adequate with the requests of our century, (2) to actively contribute to mature the hydrogen and fuel cell technology, especial competitiveness, and price, and (3) the public policy support, especially money investment.

In the anterior paragraphs, we mentioned three hubs regarding the research, development, demonstration, or introduction on the (early) market of hydrogen and fuel cell technologies: EU, US-Canada, and Japan-South Korea. What those hubs have in common? According to A. Moreno's statements in opening of the EFC15 conference, "before to be energy vector hydrogen will be a democracy vector" [22].

References

[1] Pierre-Etienne Franc, Pascal Matteo, Hydrogen: the energy transition in the making!, Editions Gallimard, Paris, 2015.
[2] European Union, Energy Research in Europe, A guide to European, national and international energy research programmes and organizations, Publication Office of the European Union, 2013.
[3] European Communities, European funded research on Hydrogen and Fuel Cells review assessment future outlook, Office for Official Publications of the European Communities, 2008.
[4] FCH JU, Multi-Annual Work Program 2014–2020, 2014.
[5] FCH JU, Trends in investments, jobs and turnover in the Fuel cells and Hydrogen sector, February 2013.
[6] FCH JU, Brochure FCH JU, produced on the occasion of the Stakeholders General Assembly 2013 (ID 196297), November 2013.
[7] FCH JU, 1st Meeting of the States Representatives Group of the FCH 2 JU, November 2014.
[8] European Communities, European Fuel Cell and Hydrogen Projects 1999–2002, Office for Official Publications of the European Communities, 2003.
[9] European Communities, European Fuel Cell and Hydrogen Projects 2002–2006, Office for Official Publications of the European Communities, 2006.
[10] European Commission, Council Regulation (EU) No. 559/2014, establishing the Fuel Cells and Hydrogen 2 Joint Undertaking, Official Journal of the European Union, L 169/108, 7.6.2014.
[11] FCH JU, Programme Review report 2014, Publications Office of the European Union, 2014.
[12] NEW-IG, Fuel Cell and Hydrogen technologies in Europe: Financial and technology outlook on the European sector ambitions 2014–2020, 2013.
[13] IEA, Technology roadmap: hydrogen and fuel cell, Paris, June 2015.
[14] Bertuccioli L., Chan A. et al., Study on development of water electrolysis in the EU, Fuel Cells and Hydrogen Joint Undertaking, February 2014.
[15] FCH JU, Trends in investments, jobs and turnover in the Fuel cells and Hydrogen sector, February 2013.
[16] E4tech, The Fuel Cell Industry Review 2015, available at www.FuelCell IndustryReview.com, November 2015.
[17] DOE, Historical Fuel Cell and Hydrogen Budgets, DOE Hydrogen and Fuel Cell Program Record, May 2015.

[18] Steven Stoft and César Dopazo, R&D programs for hydrogen: US and EU, in Security of Energy Supply in Europe Natural Gas, Nuclear and Hydrogen, Eds. François Lévêque, Jean-Michel Glachant, Julián Barquín, Christian von Hirschhausen, Franziska Holz and William J. Nuttall, Edward Elgar Publishing Limited, 2010.

[19] NEDO, Hydrogen Society has come: Beginning of the New Age of Hydrogen, Focus NEDO, No. 57, pp. 04–11, September, 2015.

[20] Robert Rose, ENE-FARM installed 120,000 residential fuel cell units, FuelCellsWorks, September 23, 2015, (available at www.fuelcellsworks.com, accessed at 27/02/2016).

[21] Marjolein de Best-Waldhober, Ruben Peuchen, Marcel Weeda, "Public perception on hydrogen storage", Deliverable No. 5.2., HyUnder (Grant agreement no. 303417), 14.10.2013.

[22] Angelo Moreno, Plenary Session Opening, Piero Lunghi Conference EFC15, Naples, Italy, 16–18 December, 2015.

Index

About the Editor

Iordache Ioan was born in Iasi Romania in 1976 and graduated from the "Gheorghe Asachi" Technical University of Iasi in the field of Environment Engineering in 2000 and continued and finalized the master's degree within the specialization of Environment Management in 2001. He completed a Ph.D. in the field of chemistry at the same university in 2007 and a second in University "Politehnica" of Bucharest in the field of Industrial Engineering in 2015. He was selected as delegate of Romania to the FCH States Representatives Group of the Fuel Cells and Hydrogen Joint Undertaking (FCH JU) in 2010 and was elected as Vice-Chair of this group in 2015. He is founding member and Executive Director of Romanian Association for Hydrogen Energy.

In the last years, his qualification and experience was focused in the field of the fuel cells and hydrogen, application and research. After a serious background in the field of sustainable development and environmental protection, including both academic and industry experiences, he was selected to work in Romanian National Center for Hydrogen and Fuel Cells. He has expert agreement with Romanian National Authority for Scientific Research and Innovation since 2010. Dr. Iordache has authored and coauthored a number of research papers and books in his field; his scientifical contribution includes also participation in conferences, participation in projects, being a member of the conference committees, becoming a reviewer, and being a member of the scientific and technical associations.

Lightning Source UK Ltd.
Milton Keynes UK
UKOW06n0633020217
293370UK00001B/28/P

9 788793 379985